AF718103

Planning Ship Replacement in the Containerization Era

Planning Ship Replacement in the Containerization Era

Henry S. Marcus
Massachusetts Institute of Technology

Lexington Books
D. C. Heath and Company
Lexington, Massachusetts
Toronto London

Library of Congress Cataloging in Publication Data

Marcus, Henry S
Planning ship replacement in the containerization era.

Based on the author's thesis, Harvard, 1970.
Bibliography: p.
1. Merchant marine—United States. 2. Shipbuilding subsidies—United States. I. Title.
HE745.M36 387.5'44 73-11662
ISBN 0-669-88534-7

Copyright © 1974 by D. C. Heath and Company

All rights reserved. No part of this publication may be reproduced or transmitted in any form or by any means, electronic or mechanical, including photocopy, recording, or any information storage or retrieval system, without permission in writing from the publisher.

Published simultaneously in Canada.

Printed in the United States of America.

International Standard Book Number: 0-669-88534-7

Library of Congress Catalog Card Number: 73-11662

Contents

List of Figures

List of Tables

Preface

This book examines the decision making process in liner replacement with emphasis on the time period 1961 through 1969. During this period the container revolution with its accompanying intermodal aspects began to develop at a fast pace. This changing technology called for new skills in planning. The tables of data in the book are generally not extended beyond 1970 in order to show the reader only those data known to the steamship companies analyzed.

Benefitting from errors made by steamship lines in 1961 to 1969, the author proposes a decision making process for planning liner replacement in the future. The Merchant Marine Act of 1970, which is described in Chapter 7, has changed some of the constraints on the subsidized steamship companies. However, these regulatory changes do not affect the validity of the author's proposed planning process; they simply modify some of the many factors which must be considered within such a process.

Acknowledgments

This book is based on my doctoral dissertation at Harvard University, Graduate School of Business Administration, submitted April 27, 1970.

The research for the study began nearly six years ago and was inspired by Professor Paul W. Cherington. He was invaluable in providing initial guidance; and throughout my work he gave of his time both while at Harvard Business School and during his leave of absence at the Department of Transportation. Professor Ram Charan helped me greatly in the structuring of the research and gave very useful comments during the writing of the manuscript. I am deeply indepted to Professor Paul O. Roberts (now at the Massachusetts Institute of Technology), the chairman of my thesis committee, who provided valuable suggestions in the earlier stages of this study and worked very closely with me throughout the writing, greatly aiding the organization of the dissertation. He bore the brunt of reading all my early drafts and showed an extraordinary amount of patience throughout my many revisions.

The list of persons from the maritime industry and from government who have aided me with advice, information and material is unfortunately too long to catalogue here specifically; however, I will list the companies involved: American Export Isbrandtsen Lines, American Mail Line, American President Lines, Delta Steamship Lines, Farrell Lines, Grace Lines, Lykes Brothers Steamship Co., Moore-McCormack Lines, Pacific Far East Lines, Prudential Lines, States Steamship Co., United States Lines, Sea-Land Service, and Matson Navigation Co. Various agencies of the Maritime Administration have also been very helpful to me. Of course, the interest taken by all these persons in my work does not necessarily indicate that they concur with the ideas expressed here.

Many persons have provided secretarial aid to me during the writing of this manuscript, but I would like to thank particularly Mrs. Jeanne Dellas and Mrs. Carole Glasser who did the majority of the typing. I am also indebted to Mrs. Karen Reixach for her valuable editorial comments. Finally, I must thank my wife for her encouragement, understanding, patience, and willingness to discuss the study. She has kindly tolerated my turning most of our small apartment into my private study for many months.

I, of course, assume full responsibility for the conclusions reached here and for errors both of commission and omission.

Glossary of Terms

Bale cubic –	The cubic capacity of a cargo space. For a break-bulk cargo hold, the volume is measured from the inside of the frames or cargo battens. For a space where cargo is loaded into containers or lighters, the interior volume of the containers or lighters is used.
"C" – type –	A designation for cargo ships, generally bearing the following relation to length: C–1, 400 ft.; C–2, 400 to 450 ft.; C–3, 450 to 500 ft.; C–4, 500 to 550 ft. Higher "C" types denote ships of greater length.
Deadweight –	The total weight in long tons (2240 lbs.) that a ship carries on a specified draft including cargo, fuel, water in tanks, stores, baggage, passengers, crew and their effects, but excluding the water in the boilers.
Light ship tons –	The total weight in long tons of the ship including the complete hull, the machinery, and all the equipment, but excluding the cargo and the other components of deadweight tonnage.
Long ton –	A measure of weight equaling 2240 lbs.
Measurement ton –	A measure of space equaling 40 cubic feet.
Short ton –	A measure of weight equaling 2000 lbs.

1 Introduction

The U.S. steamship industry is undergoing major changes in its fleet and mode of operation. The recent need for replacement of large numbers of ships has coincided with the revolution in cargo handling brought about by unitization in the form of containerships and barge-carrying vessels.

In the decade 1956–1966, the United States was losing its most valuable type of cargo to foreign competition. Of crucial importance in this situation are choices in ships that the subsidized liner operators made. Although commercial liner cargo carried in U.S. oceanborne bottoms accounts for only about 14 percent of the oceanborne commercial tonnage carried in the U.S. foreign trade, it represents approximately 70 percent of the dollar value as shown in Table 1–1. The market share of this valuable cargo under the U.S. flag decreased from 38.7 percent in 1956 to 22.2 percent in 1967, as shown in Table 1–2.

In 1967, subsidized ships carried 85 percent of the oceanborne commercial and government-sponsored nondefense general cargo moved in U.S. liners as shown in Table 1–3. Consequently, the effectiveness of this fleet will help determine the future of the U.S.-flag's share of liner cargo.

At the level of the individual firm, the market shares and financial success are being affected by the impact of unitization. Subsidized lines are being hard pressed by not just foreigners but the unsubsidized U.S. operators, who have been first to innovate in both technology and management methods. For example, Sea-Land, the unsubsidized line that pioneered large-scale containerization, entered the North Atlantic trade in 1966. In 1968, this company carried 366,000 long tons, or 52 percent of the U.S.-flag container tonnage on this trade route.[a] At the same time the subsidized lines were fighting for survival as they were being called upon to make substantial investments in the replacement of their equipment.

Careful planning is imperative when so much is at stake. In a transitional era, when a new form of technology with new economic potential is taking hold, the planning process becomes far more complicated than in the past when continuity was assured. Moreover, the investments required by unitized cargo operations are far greater, and the systems are more complex.

---

[a]Sea-Land's carryings in 1968 represented 6.4 percent of all the North Atlantic general cargo, or more than half the 12.4 percent carried by U.S. Lines, historically the dominant U.S.-flag steamship company in this trade. Data taken from "Testimony of James R. Barker on Behalf of American President Lines, Ltd. and American Mail Line, Ltd. before the Federal Maritime Commission," Docket No. 69–56, April 1970, pp. 16–17.

Table 1-1
U.S. Oceanborne Foreign Trade: Commercial Cargo Carried* Tonnage (in millions of long tons) and Dollar Value (in $ billions)

Calendar Year	*1963*	*1964*	*1965*	*1966*	*1967*	*1968*	*1969*	*1970*
Total Long Tons	311.6	332.8	371.3	392.3	387.6	418.6	426.1	473.2
Percentage of Total	100	100	100	100	100	100	100	100
Liner Tons	48.9	50.3	49.2	49.9	47.9	46.1	41.0	50.4
Percentage of Total	16	15	13	13	12	11	10	11
Nonliner Tons	136.2	161.4	171.6	189.5	190.4	209.5	211.6	240.7
Percentage of Total	44	48	46	48	49	50	50	51
Tanker Tons	126.5	121.1	150.5	152.8	149.3	163.1	173.5	182.1
Percentage of Total	40	37	41	39	39	39	41	38
Total Value	27.5	30.0	32.4	36.4	36.6	41.1	41.7	49.7
Percentage of Total	100	100	100	100	100	100	100	100
Liner Value	19.5	21.3	22.3	24.8	24.8	26.8	27.0	33.5
Percentage of Total	71	71	69	68	68	65	65	67
Nonliner Value	5.2	5.9	6.6	8.2	8.6	10.8	11.0	12.2
Percentage of Total	19	20	20	23	24	26	26	25
Tanker Value	2.8	2.8	3.5	3.4	3.2	3.4	3.6	4.0
Percentage of Total	10	9	11	9	8	8	9	8

*Includes government-sponsored cargo; excludes Department of Defense cargo and U.S./Canada translakes cargo.

Source: Derived from U.S. Department of Commerce, "The Annual Report of the Maritime Administration for Fiscal Year 1972" (Washington, D.C.: Government Printing Office, 1972), pp. 90-91.

Table 1-2
Market Share of Liner Tonnage under U.S. Flag in the U.S. Oceanborne Foreign Trade*

Calendar Year	*Market Share*
1956	38.7
1957	38.0
1958	32.3
1959	28.1
1960	28.6
1961	25.8
1962	26.2
1963	27.7
1964	28.1
1965	22.8
1966	22.9
1967	22.2
1968	24.0
1969	22.6
1970	23.5

*Includes government sponsored cargo; excludes Department of Defense cargo and U.S./Canada translakes cargo.

Source: U.S. Department of Commerce, "The Annual Report of the Maritime Administration for Fiscal Year 1972" (Washington, D.C.: Government Printing Office, 1972), p. 90.

Table 1-3
U.S.-Flag Carriage of Foreign Trade by Type of Cargo and Segment: 1967 (measurement tons in thousands)

	Commercial			*Government-Sponsored*			*DOD**	
Segment	*General Cargo*	*Bulk*	*Subtotal*	*General Cargo*	*Bulk*	*Subtotal*	*General Cargo*	*Total*
Total Liner	13,150	1,452	14,602	4,130	708	4,838	13,837	33,277
Subsidized Liner	11,178	1,016	12,194	3,511	496	4,007	6,158	22,359
Percentage of Total Liner	85	70	83	85	70	83	45	67
Unsubsidized Liner	1,972	436	2,408	619	212	831	7,679	10,918
Percentage of Total Liner	15	30	17	15	30	17	55	33
Tramp	856	3,169	4,025	794	3,310	4,104	4,705	12,834
TOTAL	14,006	4,621	18,627	4,924	4,018	8,942	18,542	46,111

*Department of Defense

Source: Derived from James R. Barker and Robert Brandwein, *The United States Merchant Marine in National Perspective,* (Lexington, Mass.: Lexington Books, D. C. Heath and Company, 1970), II-30.

In such an environment, ship replacement must become an integral part of long-range planning because the choice of ship design commits a line to a style of business as well.

Since the regulations imposed by the government intimately affect the kind of fleet and the options of each company, the current government restrictions on the fleet need to be reassessed.

The purpose of this study, then, is twofold: (1) to present a program for planning subsidized ship replacement in today's complex environment, and (2) to examine the impact of government regulation on ship replacement and recommend procedures which will enable the subsidized lines to choose their vessel designs and operate their fleets most effectively.

Existing Studies in the Maritime Field

Existing studies in the maritime field normally fall into one of three categories: (1) public policy, (2) forecasts of various factors, and (3) mathematical management tools. Public policy reports wrestle with the problem of the justification of the subsidy program and the question of how subsidies and government regulations should relate to the entire U.S. merchant marine.[b] Other publications deal with forecasting various types of cargo or predicting various technological developments.[c] Various mathematical tools to be used in planning have also been presented in many studies.[d] Such methods as computer simulation techniques have been described, and methodology for comparing costs of various methods of moving cargo have been explained. These publications have proved helpful in understanding the environment and in performing various specific analyses. However, few, if any, of these studies in the maritime field have dealt directly with the management problems of subsidized vessel replacement, and none has proposed a comprehensive method of planning in which new research findings and tools can be brought to bear on a problem in more than an ad hoc fashion.

Approach of This Study

While the U.S. merchant marine fleet is made up of tankers, bulk carriers, freighters, and passenger vessels, this study concentrates solely on freighters providing liner (scheduled) service, as distinct from tramp vessels. Emphasis is placed on the vessel replacement problems of the subsidized

[b]The three primary books in this field are Wytze Gorter, *United States Shipping Policy* (New York: Harper and Bros., 1956); Allen R. Ferguson et al., *The Economic Value of the Merchant Marine,* The Transportation Center at Northwestern University, Evanston, Ill., 1961; and Samuel A Lawrence, *United States Merchant Shipping and Politics* (Washington, D.C.: The Brookings Institution, 1966). Each is highly critical of the public programs by which subsidy is administered.

[c]These studies are described in Chapter 8.

[d]These studies are described in Chapters 4 and 8.

companies. For a subsidized liner firm, the choice of its next vessel design is a key factor in determining the future success of the company. Since each of these companies must operate within the laws of the subsidy program, government regulations are treated to the extent that they affect the firm's planning process.

Background on the environment and planning processes during the late fifties has been obtained from various publications supplemented by case studies written by the author.[e] Information on recent planning processes has been obtained by personal interviews with executives from the subsidized lines as well as from Matson and Sea-Land. In many instances this data was included in case studies written by the author. Information on the effects of government regulations on the planning and operational flexibility of the liner firms was derived by interviews in both government agencies and in industry. Again a case study was prepared containing some of these data.[f]

Planning, in this study, is defined as "the process of deciding on objectives of the organization, on changes in these objectives, on the resources used to attain these objectives, and the policies that are to govern the acquisition, use, and disposition of these resources."[g] The planning process includes the identification, data gathering, and evaluation of relevant factors.

In this study, the planning processes of four subsidized lines are examined as the firms began their new construction programs during the transition from break-bulk cargo vessels to full containerships.[h] The initial strategy of each firm is described, and the process which led to its initial design is presented. The reaction of each company is analyzed as the transition to unitization progressed. In each instance, the company was not satisfied with the products of its initial planning. The reasons for this dissatisfaction are explained. This general format is repeated to analyze the planning processes of Sea-Land and Matson, the only unsubsidized commercial liner firms to improve their fleets significantly both in the late fifties and the sixties. For each firm, the strategies used and the reactions of the firm to its ship designs are examined.

---

[e]Refer to the following cases that have been written by the author: "American Mail Line, Ltd.," 1969, "Moore-McCormack Lines, Inc., Analysis of Roll-On/Roll-Off Ships," 1969, "Executive Lines (A)," 1970, and "Sea-Line Service (A) and (B), 1970, Harvard Graduate School of Business Administration, Boston, Mass.

[f]Refer to "Maritime Subsidy Board," a case study written by the author at the Harvard Graduate School of Business Administration, Boston, Mass.

[g]Robert N. Anthony used this description to define strategic planning in his book, *Planning and Control Systems, A Framework For Analysis,* p. 16. Division of Research, Harvard Graduate School of Business Administration, Boston, Mass., 1965.

[h]Full containerships are ones that have little or no cargo space for goods which are not placed in containers. These vessels use cellular holds to store containers efficiently. Break-bulk cargo vessels are conventional ships that handle cargo on pallets, in nets, or in slings. Semicontainerships can carry both break-bulk cargo and containers. Since all break-bulk cargo vessels can carry some containers on their decks, in reality they are semicontainerships. Consequently, new vessels without cellular holds are referred to as either break-bulk cargo vessels or seimcontainerships in this study.

The decisionmaking processes which created the new barge-carrying ships are not analyzed in detail in this study because with the data made available to the author it could not be determined whether these designs were a result of good or bad planning processes.

Although many factors affect the success of a steamship line, it is imperative that adequate planning is performed for the vessel replacement process. Sufficient effort must be expended to make ship replacement an integral part of corporate strategy and long-range planning. Government regulations should also be more responsive to the needs of the subsidized lines in the changing environment.

General Organization of This Study

Chapters 2 and 3 describe the environment and the vessel-replacement planning processes of the subsidized lines from the start of the Ship Replacement Program in 1956 until 1961. The year 1961 was chosen as the end of this time period because by this year the technological and economic feasibility of full containerships had been proved in the domestic trades and studied for use in the foreign trades. In addition, by 1961 every subsidized line had planned for the construction of new vessels in the Ship Replacement Program. Chapters 4 and 5 present the changes in the subsidized environment with the advent of containerization and describe the decisionmaking processes used by four subsidized companies in acquiring new construction during this period. The environment and planning processes of Sea-Land and Matson are examined in Chapter 6. Chapter 7 compares this information to the previously presented data on the subsidized lines. Recommendations for changes in government regulations affecting the subsidized lines are also included. Chapter 8 proposes a methodology for planning subsidized vessel replacement. The conclusions of the study are presented in Chapter 9.

2 Subsidized Environment before 1961

Before 1961, the subsidized steamship companies generally provided a port-to-port service for shippers. All liner vessels were similar in their cargo handling techniques and all such vessels carried all types of cargo. The subsidy program enabled the subsidized lines to build new ships while they operated with a myriad of government regulations. The emergency ship construction program during World War II eliminated any chance of orderly vessel replacement so that when the Ship Replacement Program began in 1956 the subsidized lines had not planned for new construction for about twenty years. The stable environment placed little penalty on a subsidized line that chose a nonoptimal ship design. Consequently, there was little incentive for long-range planning.

Commercial Environment in the U.S. Foreign Trade Vessel Types

Before 1961, the general outline of liner vessels had remained the same for many years, and the basic cargo handling techniques had not significantly changed since the time of the Phoenicians.[a] Ocean-going dry cargo was carried by three types of ships: liners, tramps, and bulk carriers. The bulk carriers concentrated solely on commodities like grains or other goods which were presented in large lots and wre not susceptible to marking or counting. The liner and tramp vessels were physically similar, although the liner vessels were normally newer and faster. Both types of ships normally handled cargo in nets or slings or on pallets. Either vessel could handle all types of dry cargo.

The difference between the liners and the tramps was mainly that of service offered. The tramp operators contracted for each vessel movement and carried very large shipments, frequently shipload lots. The liner operators published sailing schedules and carried mostly small shipments. Because the tramp operator ran his vessels only when a large shipment was guaranteed, he could afford to charge a lower freight rate than the liner.

[a]A notable exception was Grace Lines' attempt at starting a full containership service in the U.S. foreign trade in 1960. This subsidized line met with immediate failure when Venezuelan longshoremen refused to handle the containers. The ships were laid up and subsequently sold. No other attempt at introducing large-scale containerization in international trade was made until 1966.

Liner cargo was generally characterized by high-value goods, often highly processed items. Tramp cargo was normally identified by most of the four following characteristics: (1) low value, which demands inexpensive transportation (this outweighs the advantage of speed and regularity of sailing); (2) great bulk or weight, or goods which steamship companies in the liner trade are neither interested in nor equipped for, such as chemicals; (3) simple handling, which requires no special facilities for loading or discharging; and (4) location in remote ports, which are out of the range of scheduled liner service.[1]

Port-to-Port Service

Historically, each mode of transportation has been concerned only with the portion of the origin-to-destination cargo movement that it controls. Consequently, the steamship lines concentrated on providing a port-to-port service, leaving the inland movement of cargoes to other industries. Therefore, the shipper or freight forwarder was responsible for planning the origin-to-destination movement of cargo shipments.

During this period, the government subsidy program enabled the subsidized lines to build new vessels. The unsubsidized companies relied on war-built vessels rather than attempt to obtain financing for construction in the high-priced U.S. shipyards.

The Government Subsidy Program

The beginning of U.S. Government assistance to the maritime industry can be traced back to 1789 when Congress prohibited documentation of foreign-built vessels. For many years indirect subsidies such as mail contracts were used. In 1935, President Roosevelt recommended outright subsidies to the maritime industry. He stated:

> If the Congress decides that it will maintain a reasonably adequate American merchant marine I believe that it can well afford honestly to call a subsidy by its right name.
>
> Approached in this way a subsidy amounts to a comparatively simple thing. It must be based upon providing for American shipping, government aid to make up the differential between American and foreign shipping costs. It should cover first, the difference in the cost of building ships; second, the difference in the cost of operating ships; and, finally, it should take into consideration the liberal subsidies that many foreign governments provide for their shipping. Only by meeting this three-fold differential can we expect to maintain a reasonable place in ocean commerce for ships flying the American flag, and at the same time maintain American standards.[2]

The philosophy of the President's statement was encompassed in the Merchant Marine Act of 1936. The act outlined the nation's maritime policy in forceful and succinct language:

> It is necessary for the national defense and development of its foreign and domestic commerce that the United States shall have a merchant marine
>
> (A) sufficient to carry its domestic water-borne commerce and a substantial portion of the water-borne export and import foreign commerce of the United States and to provide shipping service on all routes essential for maintaining the flow of such domestic and foreign water-borne commerce at all times,
>
> (B) capable of serving as a naval and military auxiliary in time of war or national emergency,
>
> (C) owned and operated under the United States flag by citizens of the United States insofar as may be practicable, and
>
> (D) composed of the best-equipped, safest, and most suitable types of vessels, constructed in the United States and manned with a trained and efficient citizen personnel.[3]

The government program provided that subsidized liner firms would receive operating differential subsidy and construction differential subsidy so that they could operate and build ships on an equal cost basis with their foreign competitors. Another benefit of the subsidy program allowed the subsidized lines to make voluntary deposits into special reserve funds. This money was tax exempt if used for ship construction, conversion, or repair, as well as lighters or containers.

All U.S.-flag vessels, both unsubsidized and subsidized, could take advantage of other government benefits such as construction loans and mortgage insurance. In addition, the government provided training for officers through its support of Merchant Marine Academies and made research and development grants for studies to improve ship design. By the Cargo Preference Act at least 50 percent of ocean transportation of government-financed cargoes was reserved for American operators but not necessarily liner operators. All Department of Defense cargo was also reserved for U.S.-flag ships.

To obtain the advantages of the subsidy program, liner firms were required to follow the following government regulations:[4]

1. The service must be determined to be essential and regular.
2. The operation must be required to meet foreign-flag competition and to promote the foreign commerce of the United States.
3. The applicant must own or be willing and able to obtain vessels of the size, type, speed, and number, and with proper equipment to meet competitive conditions and to promote foreign commerce.

4. The operator must possess the ability, financial resources, and other qualifications necessary to enable him to carry on a successful operation.
5. Moreover, no subsidy may be granted if it would give undue competitive advantage or be unduly prejudicial to citizens of the United States.
6. Subsidized freighters must be manned by U.S. citizens.
7. Only vessels built and registered in the United States can be subsidized.
8. One half of the profits in excess of 10 percent of capital necessarily employed are recaptured by the Maritime Administration up to the amount of the subsidy paid the operator.
9. Other financial conditions are imposed upon the subsidized operators. These include dividend restriction to 10 percent of capital necessarily employed and annual deposits of funds equalling the annual fleet depreciation into a Capital Reserve Fund to be used primarily for new ship construction.
10. The general operation and character of the business are also closely regulated. Thus:
    a. Subsidized lines cannot operate any charter vessels under subsidy except during a period of actual emergency.
    b. Subsidized lines are not allowed to engage in the sale of any services to the subsidized portion of the operation ancillary to the operation of the subsidized vessels. This includes such activities as tow boat services, stevedoring, and ship repairing.
    c. Subsidized lines cannot operate unsubsidized vessels in competition with other subsidized lines.
    d. Subsidized lines cannot engage in intercoastal or coastal trades of the United States.
    e. Subsidized lines are restricted from engaging directly or indirectly in any enterprise not connected with shipping.
    f. Subsidized lines cannot operate any foreign flag vessels competing with a U.S.-flag vessel on a line deemed to be essential.
    g. The contractor must operate his vessel in the most economical and manner, with due regard to wage, manning scales, and working conditions prescribed by the Maritime Administration.
11. With the signing of up to twenty years for Operating Differential Subsidy, the subsidized lines agreed to construct new vessels approximately at the time when their present ships reached the end of their theoretically economic life.[b]

As the new maritime law became effective, only nine of the forty-four companies which had received government aid through mail contracts were immediately willing to assume the obligations that accompanied operating differential subsidy agreements. Some companies preferred not to divest themselves of foreign affiliations as required by the act. Others did not wish to

[b]The economic life of the ship was assumed to be twenty years if the vessel was built before January 1, 1946; otherwise, twenty-five years.

operate under government regulation. However, the principle reason for their reluctance was the heavy investments which would be required by their obligation to build new ships.[5]

The orderly replacement of vessels was terminated by the crises leading up to World War II. At the height of the emergency shipbuilding program for the war effort, yards in the United States that in 1935 had been bare were turning out ships at a phenomenal rate, as shown in Table 2-1.[6]

Table 2-1
Ship Construction 1939–1945

Calendar Year	*Liberty Ships*	*Other Vessels*	*Total*	*Deadweight Tonnage*
1939	0	28	28	341,219
1940	0	54	54	638,037
1941	2	101	103	1,159,765
1942	542	218	760	8,091,833
1943	1,232	717	1,949	19,270,985
1944	727	1,059	1,786	16,960,985
1945 (to June 30)	84	613	697	6,960,138
TOTALS	2,587	2,790	5,377	52,761,723

Source: Joseph G. Barkan, "Spotlight on the Construction Differential Subsidy Program," The Metropolitan Section of the Society of Naval Architects and Marine Engineers, March 19, 1963. p. 8. Published herein by permission of the aforementioned Society.

This construction program frustrated one of the major purposes of the 1936 Act by postponing the establishment of a phased program of merchant ship construction. The government could hardly continue to provide funds for building vessels when it had a merchant fleet of 5,377 ships, all less than seven years old. Many of these ships were the modern and efficient C-type freighters which had no peers on the seas at that time.

After the war during the ship's sales program of 1946 to 1951 the American shipping industry purchased 847 ships from the war-built fleet. This enabled the subsidized firms to acquire entire fleets of modern vessels, but also meant that all these would later become obsolete at the same time.

The first postwar freighter to be built was the *Schuyler Otis Bland,* an experimental ship financed by the government. In 1950, the Maritime Administration requested, and Congress appropriated, 350 million dollars to construct thirty-five Mariners. A number of the experimental features of the *Bland* were incorporated into the design of the Mariner, the first new class of cargo ship to be constructed in the United States since the war. The Mariners first entered service during the latter part of 1952. Since the cargo carriers in the government's reserve fleet were frozen when the ship's sales program terminated on January 15, 1951, these ships were the only ones that could be sold to private operators. Twenty-nine of these ships were sold at a price of approximately five million dollars each to the following private operators: American Banner Lines (1); American President Lines, Ltd. (8); The Oceanic Steamship Company (2); Pacific Far East Lines, Inc. (7); Pacific Transport Lines (2); and United States Lines Company (9).

In 1956, the nation's subsidized lines, in cooperation with the U.S. government, embarked on the largest peacetime merchant shipbuilding program in the country's history. The purpose of the program was to replace the subsidized fleet of more than 300 privately owned dry cargo ships. Table 2-2 shows the vessels to be replaced; most of these ships were built during World War II and were approaching obsolescence. Their replacement was part of the government's long-range maritime policy of assuring the nation an efficient fleet of merchant vessels and a shipbuilding industry capable of rapid expansion during a possible wartime emergency.

A primary aim of this undertaking was to replace the cargo fleet gradually. By providing a steady source of work involving a total of millions of man-hours, the shipyards would be able to maintain a force of skilled workers which would be the nucleus for rapid expansion. Thus a stabilized shipbuilding program was scheduled so that some of the World War II vessels would be replaced before they reached their statutory age limit while others would be in service for some time after their legal age limit had expired. By 1961 all the subsidized lines had received delivery of new construction, had vessels under construction, or were planning new construction. As of April 10, 1961, eighteen subsidized vessels had been delivered and fifty-nine were under construction, as shown in Table 2-3.

During this period, the planning of the subsidized lines was aided by the stable environment in which they operated. The following description of the competitive structure of the industry explains why subsidized liner firms maintained relatively stable market shares from year to year.

Liner versus Liner

Most of the steamship lines making scheduled sailings over the majority of the world's trade routes were and still are organized into formal associations known as steamship conferences. The basic purpose of the conference is to reach common agreement among members with respect to transportation rates and other conditions of carriage of goods or of passengers. The steamship conference is, in miniature, an ocean freight cartel. In 1961, there were 105 active steamship conferences operating over routes directly involving U.S. foreign commerce.[7]

Table 2-2
Long Range Program of Ships to Be Replaced by the Late Seventies

Operator	*Date of ODS Contract*	*Freight*	*Passenger*	*Combination*	*Total*	*Estimated Cost* (millions)*
American Export Lines	1960	26/28	3	2/0	31	$ 507.8
American Mail Line	1958	8			8	97.6
American Pres. Lines	1954	19	2/3	2/4	23/26	522.3
Bloomfield Steamship	1954	4			4	38.0
Farrell Lines	1957	14			14	152.6
Grace Line	1956	20	2	9	31	371.9
Gulf & So. American	1958	5			5	47.5
Lykes Bros. Steamship	1957	53			53	500.3
Mississippi Shipping	1957	10/13		3/0	13	147.5
Moore-McCormack Lines	1955	43	2		45	489.2
Oceanic Steamship	1956	3			3	41.7
Pacific Far East Line	1958	9			9	127.9
Prudential Steamship	1960	5			5	47.5
States Steamship	1957	13			13	154.0
United States Lines	1959	46	2		48	679.6
Total		278/283	11/12	16/13	305/308	$3,925.4

*Based on late 1960 prices.

Source: U.S. House, Committee on the Judiciary, "Ocean Freight Industry." Prepared by Antitrust Subcommittee, 87th Cong., 2nd Sess., March 1, 1962, p. 36.

Table 2-3
Status of the Ship Replacement Program as of April 10, 1961

	Delivered		*Building*	
Operator	*Combination*	*Freight*	*Combination*	*Freight*
American Export Lines		4		8
American Mail Line				3
American President Lines				2
Bloomfield Steamship				
Farrell Lines				6
Grace Line	2		3	
Gulf & So. American Steamship				
Lykes Bros. Steamship		5		12
Mississippi Shipping		1		2
Moore-McCormack Lines	2	4		4
Oceanic Steamship				
Pacific Far East Line				2
Prudential Steamship				
States Steamship				6
United States Lines				11
Total	4	14	3	56
	18		59	
Cost in millions*	$241.1		$622.9	

*Based on late 1960 prices.
Source: U.S. House Judiciary Committee, "Ocean Freight Industry," p. 36.

The shipper had little to differentiate between liners of various companies.[c] Because of the conference system, the liner firms charged the same freight rates. In addition, they all provided the same port-to-port service. In some cases, speed was thought to have a marketing advantage for obtaining more or higher value cargo. However, according to the Northwestern study this supposed advantage was largely illusory.[8] It is possible that an advantage of a slight difference in liner speed was eliminated, in the eyes of the shipper, by possible delays occurring in ports or in the inland cargo movement. Consequently, the liner firm could not even use speed in many cases to differentiate its services from those of its competitors.

Pooling Agreements

Pooling agreements arrange for the pooling or sharing of cargo, revenue, or sailings among a number of steamship companies on a particular trade route. In 1961, there were twenty-four active pooling agreements among steamship lines in the foreign commerce of the United States.[9] Six subsidized lines belonged to at least one of these agreements.[10] The pooling agreement could be aimed at a specific commodity or the entire liner trade. As long as a steamship line possessed an adequate tonnage or an adequate number of

[c]Shippers sometimes favored using steamship firms of their own country. However, there is no data confirming to what extent this practice was followed.

vessels so that the company was not forced to turn cargo away simply because of a lack of carrying capacity, the pooling agreements helped to stabilize the firm's market share from year to year.

Liner versus Tramp

While the conference system, followed by almost all liner firms, stabilized rates among its members, it also provided a means for the shipper to differentiate between the liner and the tramp operators through use of the dual rate system. In this system, shippers who used only conference vessels obtained a 10 to 20 percent discount. A congressional report revealed that the dual rate system was "firmly entrenched in most trades."[11] Shippers of small quantities of high value cargo would typically use conference vessels on a regular basis. Infrequent shippers, or ones with large quantities of cargo, might choose a tramp vessel, particularly if rapid delivery was not critical. If the dual rate system existed on the trade routes of these shippers, they might avoid using conference vessels, or be powerful enough to ignore the dual rate laws without being penalized.

A recent Harbridge House report stated that the U.S. shipper frequently could not differentiate between the services offered by liners and tramps.[12] It is logical to assume that in some instances this could be true. A slow tramp vessel sailing directly between two ports might have had a quicker delivery time than a much faster liner using a circuitous route. In addition, although liners were supposed to provide regular dependable service, scheduled changes could eliminate this quality, as exemplified in Table 2-4. Other factors which could eliminate differences in vessel speed were port congestion, wildcat longshoremen strikes, delays in the terminals for pick-up or delivery, and inefficient inland transportation.

Table 2–4
Change in Schedule of *Santa Anita,* Grace Line, 1959

Schedule as of:	*Listed Sailing from Guatemala*	*Listed Arrival at Los Angeles*
Jan. 9	Mar. 3	Mar. 11
Jan. 16	Feb. 22	Mar. 2
Jan. 23	Feb. 27	Mar. 6
Jan. 30	Mar. 3	Mar. 10
Feb. 6	Mar. 5	Mar. 12
Feb. 13	Mar. 7	Mar. 14
Feb. 20	Mar. 6	Mar. 13
Mar. 6	Mar. 8	Mar. 15

Source: U.S. House Judiciary Committee, "Ocean Freight Industry," p. 206.

The Northwestern study showed that many commodities were regularly carried by both liner and tramp vessels. On the other hand, the Northwestern study reports a long list of comodities for which liners carried over 99 percent of the tonnage inbound or outbound on many routes.[13] Consequently, the tramp operators did not greatly upset the stable market shares of the liner firms.

Economies of Scale

The stability of the market shares of liner firms of all sizes was aided by the fact that a large firm had little or no advantage over a small firm during this period of break-bulk cargo operations. The Northwestern study analyzed the following possibilities for economies of scale in liner firms:

a. exploitation of the rather clear-cut economies of large ships and clustering of ports;
b. more efficiency in the use of shore facilities; and
c. economies to customers through regularity and frequency of service.

They learned that the first type of economy could be achieved by very small firms. Possibly the second type could be continued indefinitely as size increased; there was no direct evidence on this point. The third might, in very small markets, lead to increased load factors for operators rendering regular and frequent service, provided there was a net advantage to the shipper in dealing with one rather than with several shippers. The Northwestern study concluded that "even if large scale operations in a single market do permit realizing economies of all three types, analysis indicates that the economies are likely to be very small and to become exhausted at rather small sizes of firm."[14]

Liner Industry Success

Since market shares between liner firms would remain relatively stable as long as companies possessed adequate tonnage, the success of each liner firm did not depend on its performance relative to the other liner companies as much as on the general supply and demand characteristics of cargo space, which affected its competition with tramp operators. As William T. Moore, Chairman of the Board and President of a subsidized firm, Moore-McCormack Lines, Inc., stated concerning the steamship business: "it is a business that has always thrived on someone else's bad luck. By this I mean that when you examine the rates the steamship companies charge, you will see that as soon as somebody begins a shooting conflict, rates go up, and when everything is normal, they go down."[15]

When a Congressional subcommittee explained why U.S.-flag ships accounted for only 29.8 percent of the U.S. commercial oceanborne foreign trade carried in liner service in 1959, as compared to 37.6 percent five years earlier and 49 percent in 1951, it did not point out differences in services provided, but stated that perhaps the most important factor had been the failure of American shipbuilding to keep pace with foreign construction.[16] Table 2-5 shows that the percentage of active U.S. freighters in the world freighter fleet decreased from 9.1 in December 31, 1951, to 5.0 in December 31, 1958.

Table 2-5
United States and Maritime World Freighter Fleets*

	World Fleet	*U.S. Active Fleet*	*U.S. Active Fleet as a Percentage of World Fleet*
September 1, 1939			
Number of vessels	9,289	608	6.5
Deadweight tons	53,756,000	4,539,000	8.4
June 30, 1946			
Number of vessels	9,476	1,833	19.3
Deadweight tons	70,900,000	18,111,000	25.5
December 31, 1951			
Number of vessels	10,063	919	9.1
Deadweight tons	74,267,000	9,613,000	13.0
December 31, 1958			
Number of vessels	12,462	620	5.0
Deadweight tons	96,218,000	6,614,000	6.9

*Excludes vessels on the Inland Waterways, Great Lakes, and those owned by the U.S. Army and Navy and special types such as cable ships, tugs, etc. Includes freighters, refrigerated-freighters, and bulk-carriers in foreign, coastwise, intercoastal, and noncontiguous trades.

Source: Derived from Ferguson et al., *The Economic Value of the United States Merchant Marine,* The Transportation Center at Northwestern University, Evanston, Ill., 1961, p. 26.

Long-Range Planning

The stable environment simplified the planning processes of the subsidized lines. Even if a firm constructed a ship of a suboptimal design, the competitive structure of the industry would ensure a stable market share for the subsidized line with adequate tonnaging. Consequently, there was a low penalty cost involved with designing a nonoptimal ship. The combination of a stable environment and little danger of economic penalties from a vessel replacement planning process resulted in little need for long-range planning. The traditional decision-making process used in subsidized vessel replacement is described in Chapter 3. The stabilized break-bulk cargo environment is summarized diagrammatically in Figure 2-1.

Pier-To-Pier Cargo Movement

Generalized Ship Design and Terminal Facilities

↓

Shipper Cannot Differentiate Among Competing Liner Services

↓

Relatively Stabilized Market Share

↓

Low Penalty Cost of Designing Nonoptimal Ship

↓

Little Need for Long-Range Planning

↓

Standard Traditional Ship Replacement Planning Process

Figure 2-1. Stabilized Break-Bulk Cargo Environment.

3 Subsidized Ship Replacement Planning Process before 1961

Until the Ship Replacement Program started in 1956, the subsidized lines had had little opportunity to plan orderly ship replacement in their fleets for almost twenty years. Since the start of the subsidy program in 1936, decisions on ship design had been in the hands of the government. World War II vessel construction, the Ship Sales Program, and the Mariner Program had furnished the maritime industry with government-built vessels. Consequently, the subsidized lines had no permanent staffs for planning vessel replacement. The necessary planning at the start of the Ship Replacement Program was done by men who were ordinarily employed on day-to-day engineering, marine operations, or ship repair problems. Because of these other duties, ship replacement problems were sometimes left to the last minute.

Traditional Planning Process for Subsidized Vessel Replacement[a]

In theory, an economic evaluation of a ship replacement decision had to take into account all factors of revenue and cost for the life of the investment, which was twenty to twenty-five years. The major variables affecting the revenue or demand side of the picture during this period were typically: growth in world population, economic conditions in the world, wars, droughts, famines, tariffs and other trade restrictions, national pressures to start or expand various industries, increased industrialization in certain countries, and factors affecting various modes of transportation. On the cost side, there were the rising costs of longshoremen's and seamen's wages and benefits, repair work and maintenance, labor strikes, vessel depreciation, and the changing cost of fuel. Obviously, it was impossible to predict the timing of some factors such as wars, economic crises, and strikes, much less their effect on shipping operations. Consequently, shipping companies usually simplified the planning process. The major factors considered were market survey, cost calculations, vessel speed, capacity, and revenue potential.

Based on the average vessel performance for the preceding three to five years, the freight traffic department developed a rough approximation of the

[a]Parts of the description of the traditional planning process have been taken from the "Executive Lines (A)", a case study at the Harvard Business School written by the author, 1970, pp. 1–5, and the "Trade-Fleet Study" by the National Academy of Sciences–National Research Council, Maritime Cargo Transportation Conference, Washington, D.C., 1960, pp. 4–14.

amounts and proportions of cargo types which would be available on an "average" voyage. This was modified for probable changes in trade circumstances, or changes in the intentions of the company in soliciting certain types of cargo. These modifications were generally conservative. The result, then, was a listing of cargo tonnages by type of cargo available for shipment between each port-pair, for an average round-trip voyage. Current freight rates were used to generate possible revenue figures. The vessel itinerary was laid out based on past practices but was subject to modification to allow for any difference in speed of the vessel under investigation, desire by the company to improve its services, and the company's estimates of shippers' desires. Based upon the selected vessel itinerary and ship capacity, voyage revenue was computed.

On the cost side of the ledger, stevedore costs were based upon experience, or, if new handling methods were being considered, estimates were made for each port and commodity. Estimating vessel operation costs was a fine science in the opinion of planners. Given the basic characteristics of a vessel, the operating departments of most companies could compute rapidly and easily the vessel operating costs including port charges plus or minus 10 percent. Amortization of the vessel was a simple bookkeeping routine, based on estimates of vessel construction cost. These estimates were also reasonably accurate.

Because the market forecast played a key part in the traditional method, a company sometimes did research in this area. One subsidized line was part of a large trading company conversant with the foreign areas served by the line. The economic development department prepared extensive area trade and individual commodity analyses and five-year forecasts for the trading company. For the steamship line, it provided fifteen-year estimates. Other lines sometimes made use of outside consultants to aid in their cargo forecasts. However, the cargo analyses of many companies can probably be summed up by the statement of one vice-president in charge of traffic who freely admitted that he would not "bet a nickel" on the cargo forecasts he had supplied to his company's New Construction Committee.[1]

The company then began considering vessel characteristics. Although it might seem that a company could try an infinite number of combinations of speed and size for its prospective ship design, the company could actually reduce this number very simply. The speed was the more important factor to choose because the size which would produce the optimal operating characteristics could be approximated by engineering calculations once the speed was chosen.

Steamship companies normally tried to provide weekly sailings on a highly competitive trade route, based on a marketing belief that shippers could be induced to export regularly on a weekly basis. The range of voyages allowed by the government restrictions on ODS (operating differential subsidy) usually also pointed to either weekly or biweekly service.

Because conventional ships of any size spent roughly 50 percent of their time in port, a large difference in speed was required to vary the number of ships providing weekly service on a particular trade route. For example, if a company used six ships of 18 knots to provide weekly service, each ship with a 42-day round-trip voyage might spend 21 days in port and 21 days at sea. If the company wanted to replace its fleet with five ships of a higher speed, the faster ships would still spend 21 days in port but could only spend 14 days at sea in order to provide a 35-day turnaround needed to preserve weekly service. The required speed would then be 26.9 knots. In the break-bulk cargo ships, the savings from having one less ship was normally outweighed by the much higher fuel cost and the higher initial cost of the 26.9 knot ship. If the company wished to reduce its fuel cost by having a fleet of seven slower ships, the ships would have to spend 28 days at sea to preserve the weekly schedule. A speed of only 13.5 knots would be required. This speed would probably be too slow for marketing purposes to compete with 18-knot ships. In addition it was below the Maritime Administration minimum speed requirement of 18 knots. Consequently, in this instance, a steamship company would arrive at a required speed of 18 knots from just a few simple calculations. From the required speed, the size and other engineering characteristics could be derived so that from an infinite number of possible combinations of speed and size, the process was simplified to "back of an envelope" proportions.

Further variations in a vessel's size and speed could also result from management's experience and intuition. Moreover, with such an inexact science of planning, a dominant personality could have a strong effect. For example, an aggressive traffic manager might have pushed for a ship capable of high speed to meet schedules when port delays took place. Since subsidized steamship lines did not maintain full staffs of naval architects and marine engineers, they relied on outside firms to do their detailed design work. Consequently, outside naval architectural firms often influenced the vessel replacement process.

It should be stressed that the "average voyage" computations were the core of the traditional method. In general, the estimated number of voyages which would be made in an operating year was then multiplied by the "average voyage" figures to arrive at annual performance estimates.

Profit and loss estimates based on annual performance were also computed for 50 percent, 75 percent, and 100 percent load factors, and a breakeven load factor was also computed. In these computations, cargo handling costs, tonnages, and revenues were raised or lowered by a fixed percentage corres-ponding to changes in load factor.

The method of calculation was fundamentally identical to the projected voyage revenue calculations which the companies used continually in their day-to-day operations. Thus, this traditional method fit right in with existing office routine in terms of the machinery of calculations. In addition, the

Maritime Administration set out very specific instructions which required operators to present data in this form when applying for construction differential subsidy.

Influence of the Maritime Administration

In order to gain construction differential subsidy and an operating differential subsidy, a line needed the approval of several divisions of MARAD (The Maritime Administration). Consequently MARAD was a powerful influence in the planning process. When a line applied for construction differential subsidy funds, it prepared a formal presentation showing justification for its vessel replacement decisions. Because MARAD had no set of rules for judging presentations or granting funds, the elaborateness of a presentation sometimes depended on the general atmosphere at MARAD and the availability of funds. A steamship line always showed that its vessel replacement decision resulted in new ships which provided a reasonable return on investment and were more efficient than the ones they replaced.

Because ODS payments included crew wages but not the ship's fuel, it was to the government's advantage to persuade the subsidized lines to increase the speed of their new vessels. This additional speed would also prove valuable for military purposes in time of war. The Northwestern study pointed out that through government influence subsidized liners were constructed with speeds slightly higher than their economic optimum.[2] The government set a minimum of 18 knots for subsidized ships but encouraged higher speeds. In a particular instance, showing the extent of government influence, American Mail Line decided to build a flight of 18-knot ships in the late fifties; however, after more than a year of negotiations, they decided to accept the government's suggestion of constructing 21-knot vessels.

The government influenced the design of new vessels by varying the requirements for obtaining ODS as well as CDS (Construction Differential Subsidy) funds. As part of the Essential Trade Routes concept, there was not only a continually updated determination made of the number of voyages of American flag registry which were considered necessary to carry the U.S.-flag portion of the tonnage desirable, but there was also a study made of the "type, speed, and other requirements" of the tonnage which should be employed on the subsidized services.

For example, consider the following excerpt from the Essential Trade Route determination for Trade Route No. 17, relating to the trade to Indonesia-Malaya:

> Mariner-type vessels are suitable for long-range operation. C-3 type vessels are suitable, pending replacement due to age, and C-2 and Victory-type vessels are considered suitable for interim operation only,

and should be replaced at the earliest possible date. Replacement vessels should be comparable in speed, base cubic and deadweight capacity to the Mariner-type vessel with adequate deep tank and refrigerated capacity."[3]

As a general proposition, older, smaller vessels, or warbuilt Victory-type tonnage were not considered qualified for longer voyages. Since a subsidized line would not build new vessels unless it was assured of both CDS and ODS, the government's ODS requirements had a definite effect on ship design.

From the time when a subsidized line applied for CDS funds to purchase a group of ships, called a new flight of ships, until all the ships were in service normally took between four and eight years. Therefore, the company could simply perform a preliminary analysis at the time of the CDS application; then, several years, later, just before the new ships were delivered, the company could analyze the current situation in great depth to decide how best to use the new vessels. As long as the environment remained relatively stable, the subsidized lines could use this procedure to avoid the need for extensive long-range planning while initially designing their new ships. Since the analysis could be performed off-and-on over a period of years, there was no need to have personnel working full time on this task.

The Effectiveness of the Traditional Planning Process

In evaluating this process the risks of approximations and assumptions must be traded off against the costs of obtaining precise information. Most planning involves what Herbert Simon calls "satisficing." "Administrative man," wrote Simon, "satisfices–looks for a course of action that is satisfactory or good enough. Examples of satisficing criteria that are familiar enough to business men, if unfamiliar to most economists, are share of market, adequate profit, and fair price."[4] To amplify his concept, Simon presented this example:

Suppose the problem be to select a dam site for a storage reservoir. For simplicity it will be assured that the desideratum is to secure a specified volume of water storage at a minimum cost, and that water storage above the specified amount will be of no value. Usually the real problem is not so simple. The cost can be estimated, for each point along the river, of building a dam with the required storage capacity. However, to make an accurate estimate, detailed studies would need to be made of the foundation conditions at each point. Then this huge array of cost estimates could be compiled and the dam site with least cost selected.

Actually, the engineer proceeds quite differently. By inspection of a topographic map, he immediately picks out a half-dozen "plausible"

> dam sites, and forgets the rest. He is sufficiently familiar with dam construction costs to know—with a fair degree of certainty—that any other site he might choose would have a higher construction cost. Next, he makes an approximate estimate of dam costs for each of the plausible sites, assuming "normal" foundation studies as a basis for final estimates.
>
> At each step in this process there is a chance that the dam site which really is most desirable will be eliminated without complete analysis. He must exercise great skill in determining the degree of approximation that is allowable at each point in the procedure.[5]

The ship replacement planning process is analogous to the problem of selecting a dam site. The satisficing criteria used in ship replacement by subsidized lines could have been a ship design more economical than the one it replaced, one that would be approved by the Maritime Administration, and one that would be as fast or slightly faster than competitors. These criteria were met by the ship designs produced by the subsidized lines. Each new ship built was always larger and faster than the one it replaced and always at least as large or as fast as any other ship in the company's fleet.

As Simon pointed out in the satisficing process, the optimum result can be completely overlooked. However, during the period when general cargo was moved solely in conventional break-bulk cargo ships, there was little penalty in choosing a nonoptimal design. All new ships were simply modifications of two or three basic designs. Shippers could not differentiate between services of various liner firms, and the conference agreements helped to ensure that each subsidized firm would retain its market share. Consequently, if a liner firm simply replaced its vessels with ones a little bigger and a little faster, it could not go far wrong. If the ship turned out to be too small, the firm had to turn away cargo in peak periods. If the vessel turned out to be too large, the firm had to pay a slightly higher construction and operating cost. Since steamship companies could normally interchange their various vessels on their different trade routes, a line might be flexible enough to eliminate the penalty of a suboptimal design by juggling its ships around. As long as the company maintained its market share, the penalties for choosing a suboptimal ship design were relatively small, and it was not obvious that the cost of extensive long-range planning was worthwhile. Hence, the environment allowed a wide tolerance of accuracy in planning processes and thus pressured for maximum "satisficing."

The traditional process did serve the subsidized lines adequately during the break-bulk cargo era. From the start of the subsidy program in 1936 until 1960, the total profits of all U.S. subsidized lines taken together showed significant growth as illustrated in Table 3-1. However, the environment in which the subsidized lines operated began to change rather significantly after 1960. The concept of containerization began to take hold. The effect of this change on the traditional planning process is described in Chapter 4.

Table 3-1
Net Profit of Subsidized Operators
(Annual Average)

Years	*Net Profit (in Millions)*
1937-41	$30.6
42-45	28.8
46-49	49.0
50-54	41.1
55-59	61.0

Source: Samuel A. Lawrence, *United States Merchant Shipping Policies and Politics* (© 1966 by The Brookings Institution, Washington, D.C.,) p. 365.

4 Environment of the Subsidized Lines: 1961 and Beyond

By 1961, Sea-Land and Matson had proved the economic and technological feasibility of containerization in the domestic trades. Some subsidized lines had experimented earlier carrying containers as deck cargo, as shown in Table 4-1, but after Grace Lines' failure to initiate full containership service in 1960, no subsidized vessel attempted to provide full containership service until 1966. A variety of reasons, both operational and institutional, caused this reluctance on the part of the subsidized companies.

During this same period, studies were performed showing the economic feasibility of full containership service in the U.S. foreign trade. The full containership was proved to be economically superior to both semicontainerships and break-bulk cargo ships. The benefits of utilizing dockside cranes rather than shipboard cranes were also shown. In addition, this research made it evident that unitized cargo operations greatly increased the need for long-range planning in the subsidized industry. Factors which were of little significance in break-bulk cargo operations, such as inland transportation and economies of scale, were of great importance in container operations.

Reluctance of the Subsidized Lines to Adopt Large-Scale Container Operations

Many reasons were cited for the reluctance of the subsidized lines to adopt large-scale container operations: lack of cargo flow data, steamship conference objections, labor objections, lack of shipper-carrier enthusiasm, absence of expensive specialized equipment, nonstandard container sizes and fittings, archaic government regulations, and customs complications.[1]

Because containerships were required to move containers in both directions on their trade routes, the danger of an imbalance in containerizable cargo was present. This difficulty was exacerbated by a lack of cargo flow data. Even after the presence of sufficient containerizable cargo in both directions was verified, a subsidized line was hindered from innovating by the objections of its competitors. Steamship companies reluctant to make the large investment necessary to start large scale container service institutionalized this reluctance through conferences. Although conferences did have special rates for containers, these rates did not allow the shipper to save a substantial amount of money, thereby discouraging the use of containers. Labor unions also impeded the

Table 4-1
Approximate Date Containers First Handled by Subsidized Lines*

American Export Isbrandtsen Lines, Inc.	Early 1958
American Mail Line, Ltd.	March 1965
American President Lines, Ltd.	1950
Delta Steamship Lines, Inc.	March 18, 1966
Farrell Lines, Inc.	June 1966
Grace Line, Inc.	January 1960
Lykes Brothers Steamship Co., Inc.	1959
Moore-McCormack Lines, Inc.	April 1965
Pacific Far East Lines, Ltd.	1953
Prudential Lines, Inc.	1960
States Steamship Co.	1958
United States Lines, Inc.	1957

*In most cases, subsidized lines simply carried containers on the decks of break-bulk cargo ships on the date referred to here.

Source: U.S. Senate, Committee on Commerce "Standardization of Containers." Prepared by Subcommittee on Merchant Marine and Fisheries, 90th Congress, 1st sess., July 1967, p. 34.

transition to large-scale containerization. Although containers could save many man-hours of stevedoring in the ports, the longshoremen did not wish to aid this innovation without being financially rewarded.

Although shippers did not object to containerization like some labor unions, they were generally apathetic to this new form of transportation. One explanation for this lethargy was that the average manufacturer engaged in foreign trade may have been several steps removed from the actual business of transporting his cargo through a port. He may have sold through a professional export/import house. Consequently, he would have dealt solely with a freight forwarder, who, in turn, dealt with a variety of steamship companies, truckmen, railroads, warehousemen, and others.

If a subsidized line was not deterred from innovating by the above factors, it was required to make a large investment in vessels, containers, and bogies before starting a containership operation. If a line wished to own the marine terminal facilities it used, the necessary start-up costs were further increased. Any attempt to form container pools with other companies was hindered by the use of nonstandard size containers by Sea-Land and Matson, the two largest containership operators. While the standard lengths for containers were 10, 20, 30, and 40 feet, Sea-Land used 35-foot containers and Matson used 24-foot containers.

After Grace Lines' failure at providing containership service in 1960, the government did not encourage the building of full containerships during the early sixties. In addition, attempts of the U.S. subsidized lines to build new vessels were delayed by the bureaucratic constraints of the subsidy program. Time was required for various government agencies to perform economic and technological feasibility studies. A company might have also had to wait for congressional appropriation of more CDS funds if they were not available.

Moreover, U.S. shipyards lacked the machinery necessary to build ships as fast as their foreign competitors. This lack of shipyard investment in facilities was caused by the uncertainty of the government funds for construction of merchant vessels. All factors considered, a subsidized line realized that a foreign competitor could copy the new design and have his foreign-flag vessels in service before the subsidized ships were launched. This situation occurred in the sixties when Grace Lines awarded a design contract for new vessels in August 1961. Although a foreign competitor on the same trade route did not award a similar design contract until almost two years later, all four new vessels were in service before Grace Lines had received delivery of its first new ship. Details on the timing of these events are shown in Figure 4-1.

Once a company received the necessary equipment, its ability to move containers door-to-door was impeded by government regulations preventing the formulation of single-factor or through rates from one inland to an overseas inland point. Customs inspections, which have traditionally been performed on piers and in ports, also hindered the flow of containers. A container, to achieve its ultimate objective, must pass through ports to be opened initially at inland points. At the same time that the above factors were discouraging subsidized lines from starting large-scale container operations, studies were performed proving the economic benefits of such operations.

The Feasibility of Containerization in U.S. Foreign Trade

Although the subsidized lines were reluctant to start large-scale container operations, there had been sufficient evidence to show that full containerships were economically feasible in the U.S. foreign trade. Productivity could be greatly increased with full containerships. While a break-bulk vessel spent about one half of its time sitting in ports, the full containership would spend as little as one fifth of its time in port. With this increase in productivity fewer containerships would be needed to replace a fleet of break-bulk cargo vessels. The reduced turnaround time was a result of more rapid cargo handling, another source of substantial savings with containerization.

The economics of full containerships versus semicontainerships heavily favored the former. In 1959, a National Academy of Sciences report compared the oceanborne costs of moving cargo in vessels that were 0 percent, 20 percent 40 percent, 60 percent, 80 percent, and 100 percent containerized between

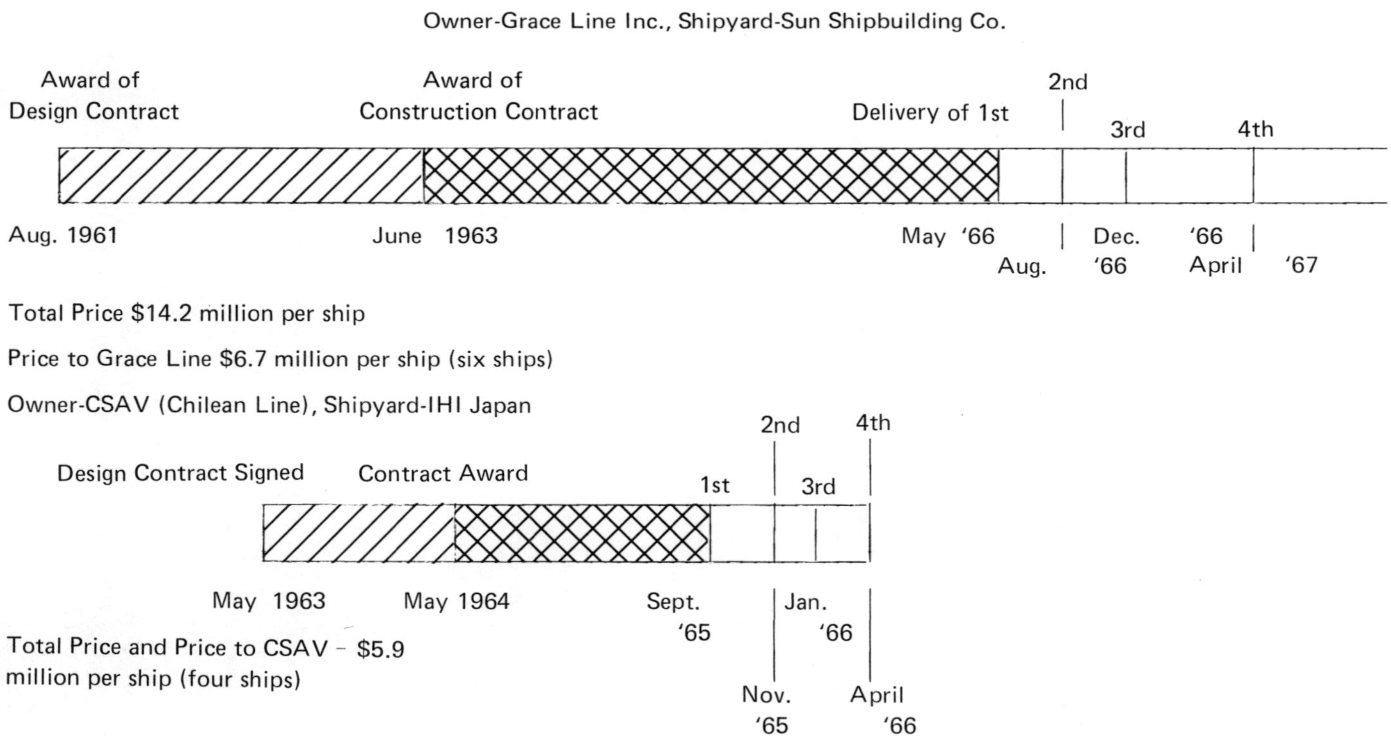

Source: U.S. Senate, Committee on Commerce, "U.S. Maritime Policy." Prepared by Subcommittee on Merchant Marine and Fisheries, 90th Congress, 1st Sess., April, May, June, and August 1967, p. 306.

Figure 4-1. Comparison of Ship Procurement Experience for Comparable Ships of Competing Steamship Companies.

two ports 5,000 miles apart. If the relative cargo handling rate (RCHR) of containerized cargo was at least four times faster than break-bulk cargo, a full containership was proved economically superior to either a semi-containership or a break-bulk cargo vessel. A sample of the summarized cost data is shown in Table 4–2.

Table 4–2
Cost per Long Ton of Moving Cargo with Various Degrees of Containerization*

Percentage of Cargo Unitized	RCHR†	*0*	*20*	*40*	*60*	*80*	*100*
Net Cost/L.T.($)	1	24.30	26.50	28.50	30.60	32.30	34.10
	2	24.30	25.00	25.40	26.00	26.80	26.70
	4	24.30	24.00	23.90	23.70	23.50	23.30
	6	24.30	23.70	23.60	22.80	22.40	21.50
	8	24.30	23.70	23.20	22.80	21.90	21.50

*This analysis assumes a distance of 5,000 miles between ports.
†RCHR is the relative cargo handling rate of containerized cargo versus break-bulk cargo. The loading rate of break-bulk cargo is assumed to be 18.75 L. T. per hour, with the unloading rate 85 percent of this amount.
Source: *Maritime Transportation of Unitized Cargo,* Publication 745, Maritime Cargo Transportation Conference, National Academy of Sciences-National Research Council, Washington, D.C. 1959.

The following year, a study by MacMillan and Westfall reported that cargo in twenty-foot containers could be handled at a rate approximately twelve times faster than break-bulk cargo.[3] Using forty-foot containers would double this rate. With these cargo handling rates, semicontainerships could not economically compete against full containerships assuming an adequate supply of containerizable cargo. In 1963, a follow-up report to the earlier National Academy of Sciences study agreed with the inferiority of semicontainerships and did not include them in their calculations. This latter report expanded the earlier study to include the cost of moving cargo from one inland to another inland point overseas.[4] Since more occasions for cargo handling were eliminated in the origin-to-destination movement of containers than in simply the ocean-borne movement, this follow-up report amplified the benefits of containerization relative to the earlier study. A sample of the cost data appears in Table 4–3.

By 1963, these publications had shown that full containerships were economically superior to either semicontainerships or break-bulk vessels in carrying containerizable cargo over such distances as found on the North Atlantic or Far East trade routes.[a] Independent research by Matson had

[a]The interport distances on trade routes 5-7-8-9, the North Atlantic, and trade route 29 in the Far East fall in the 3240-to 5000-mile range.

Table 4–3.
Total Shipper-To-Consignee System Costs For Export Cargo, Assuming Two 220-Mile Truck Line Hauls* (dollars per measurement ton)

Type of System	*Total System Cost*
Break-Bulk	$49.66
Using Containers:	
Cargo unitized at shipper's premises	
20 ft. containers	31.00
40 ft. containers	30.65
Cargo unitized at unitizing station, located within local cartage zone of shipper's premises	
20 ft. containers	32.36
40 ft. containers	32.01
Cargo unitized at marine terminal, port of origin	
20 ft. containers	31.65
40 ft. containers	31.25

*This analysis assumes a distance between port of 3240 miles and a cargo stowage factor of 76 cubic feet per ton.

Source: *Inland and Maritime Transportation of Unitized Cargo,* Publication ISBN 0-309-01135-3, Maritime Cargo Transportation Conference, National Academy of Sciences-National Research Council, Washington, D.C. 1963.

shown it that the cost of handling uncontainerized cargo as part of a containerized system was "absolutely prohibitive."[5] If break-bulk cargo was carried on a containership, the amount of port time necessary would be dictated by break-bulk cargo. Consequently, the vessel would lose the quick turnaround time and high productivity necessary to offset its high investment. If containers were carried on a break-bulk cargo vessel, the costs of the ship would increase with costs of container ownership or leasing, handling, and storing, without giving any major benefits. Therefore, semicontainerships had the disadvantages of both the containership and the break-bulk cargo vessel.

Additional research by Matson in the late fifties convinced the company that cargo handling equipment should be placed on the dock rather than on the ship because:

1. fewer cranes would be needed for a fleet operation;
2. taking the cranes off the ship increased its carrying capacity; and
3. dockside cranes could be designed to operate faster than shipboard cranes.

Sea-Land confirmed the findings of Matson's research by changing from shipboard to dockside cranes in 1965.

New Factors in Planning

Many factors in the planning environment changed with the advent of unitization. Because door-to-door service is required, origin and destination data on cargo movements are needed. More information on cargo flows is needed in order to determine the amount of containerizable cargo and the extent of any imbalance in trade.

Since the container can move inland, knowledge of inland transportation networks is required. With break-bulk cargo operations each mode of transportation in the origin to destination movement operated independently. Consequently, cargo movement was frequently delayed for several days because of the lack of coordination between modes as well as the many handlings that were necessary. With containerization, it is possible to run a container-on-flatcar (COFC) train into a port in the morning and have the containerized cargo moved to the pier and onto the vessel by nightfall.[6]

Domestic transportation carriers normally are more concerned with domestic rather than international carriage, since most of their business comes from domestic cargo movements. Consequently, this is an added incentive for steamship lines to develop expertise in the area of inland transportation. The formation of inland container pools also has to be considered.

Companies building barge-carrying ships like LASH (Lighter Aboard Ship) vessels under construction for Prudential Lines and Pacific Far East Lines and Sea Barge Clippers under construction for Lykes must also be concerned with inland transportation. These firms will provide mainly a port-to-port service with their barges; however, to the extent that they will carry containerizable general cargo rather than bulk commodities, they will eventually be competing with containership services. Therefore, to understand their market and their competitors, they must consider the inland movement of cargo even if they do not plan to provide such service.

Full containership service also brings a lowering of freight rates to reflect the lower costs of this method of transportation. Although steamship conferences had been opposed to lowering container rates, the use of full containerships outside the conferences by Sea-Land and American Export Isbrandtsen Lines has provided pressure for lower rates. Since full containerships offer better service at lower prices, shippers can differentiate between various liner services. Consequently, the previously steady market shares of the break-bulk operators have rapidly become unstable. From its initial operations on the North Atlantic in 1966, Sea-Land progressed to carrying 366,000 long tons of cargo in 1968, which represented 52 percent of the U.S.-flag participation in the container trade and about 25 percent of the total carriage of container tonnage by all flags.[7b]

[b]Sea-Land's 1968 carrying represented 6.4 percent of the North Atlantic general cargo market, more than half the 12.4 percent carried by U.S. Lines, the dominant U.S. flag carrier on the trade route.

Because different companies are now providing different services, marketing becomes important. Salesmen need retraining to cope with the problems of intermodal transportation. Additional personnel may be needed to work on inland transportation networks, long-range planning, and computerized container control systems.

Unitized-cargo operations are capital intensive rather than labor intensive as in the break-bulk cargo operations. The comparison is shown in Figure 4–2. The high fixed cost feature of unitized-cargo operations has resulted in the introduction of economies of scale.

Although economies of scale had appeared in the break-bulk cargo era in relation to size of vessel, the size of cargo liners had been limited by two factors. In the first place, it was not worthwhile to build giant ships so long as a greater part of their economic capability and capacity was lost by long periods spend in port. Secondly, the large number of competitors restricted a steamship company from getting adequate utilization of a giant vessel.[8] Both these restrictions were eliminated with unitization so that one consultant firm predicted that increasing ship capacity from 300 to 3,000 containers would reduce unit costs 42 percent on the North Atlantic trade routes.[9]

The first restriction on productivity was eliminated by the quick port turnaround time of the containerships. The second restriction on number of competitors was nullified by the economies of scale in container operations so that a large firm had an advantage over a small firm, a condition that did not exist in the break-bulk cargo era.

Economies of scale favoring large firms appeared in three areas of a container operation: (1) operation of port facilities, (2) the number of containers required in the system, and (3) the operation of a computerized container control system. Similar benefits exist in a large-scale barge or lighter operation; however, the cost advantage is smaller since the cost per unit of cargo space in a barge or lighter is less expensive than in a container.

A simple example with very rough estimates illustrates the benefits of economies of scale. Assume a small company was serving four ports on one trade route on a weekly basis with three containerships. Also assume a large firm was using twenty-seven containerships to serve twelve ports located in several major trade routes. All twelve ports were interconnected by some complex schedule of overlapping trade routes where each ship visits four ports on a twenty-one day round trip voyage as in the previous case. For simplicity in the example assume that all ships carried 500 forty-foot containers. In the case of the small company, each port received one ship per week and handled 500 containers (250 unloaded, 250 loaded). In the case of the large company each port received three ships per week and handled 1500 containers (750 unloaded, 750 loaded).[c]

[c]Since three ships made four port calls per week in the case of the smaller company, twenty-seven ships or nine times as many made thirty-six or nine times as many port calls in the case of the larger firm. Since the latter company only served twelve ports, each port received three vessels per week.

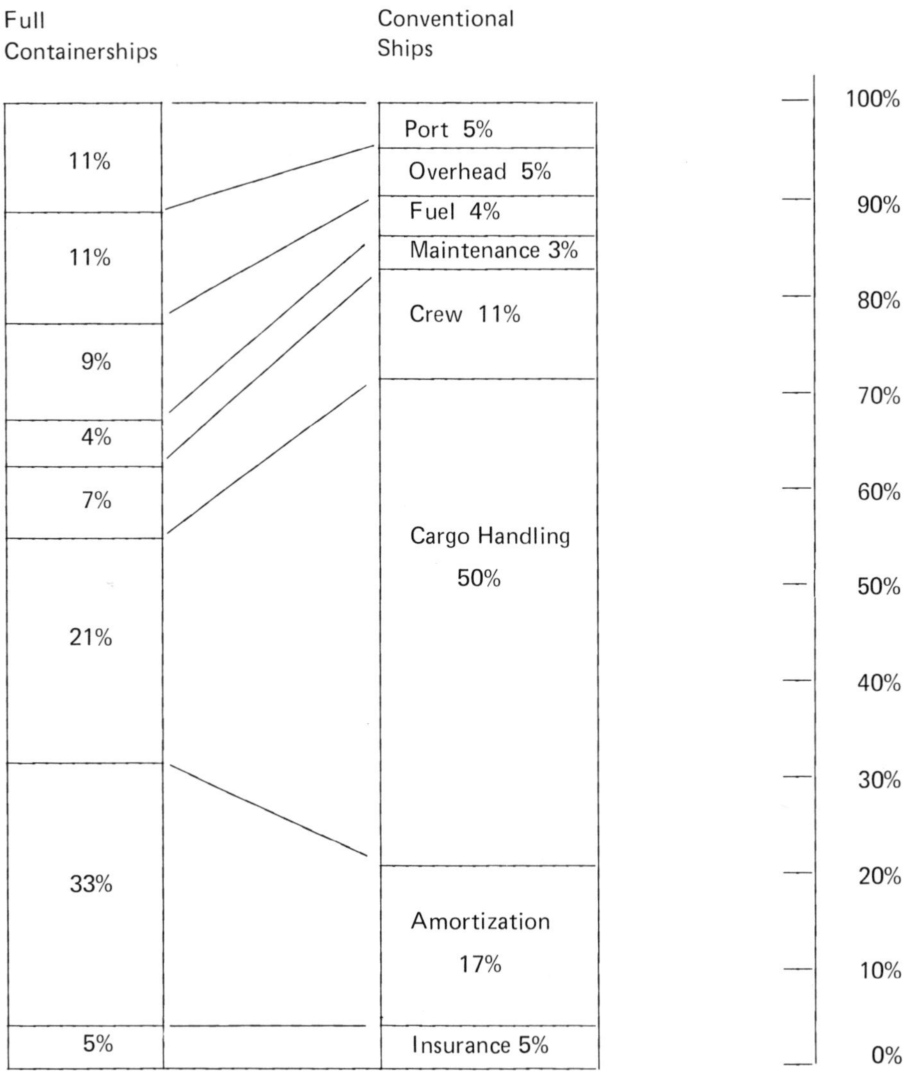

Note: The graphs show percentage operating costs for a conventional 20-knot cargo ship (Mariner) and a proposed unitized cargo, semiautomated ship of the same speed on an 8000-mile round voyage. Areas represent actual costs in dollars.

Source: Edward V. Lewis, "Research Toward More Efficient Transportation by Sea", The Society of Naval Architects, New York, N.Y., 1961 Transactions, p. 13. Published herein by permission of the aforementioned Society.

Figure 4-2. Comparative Operating Costs

Operation of Port Facilities

Since stevedoring became a capital-intensive operation, the cost per container moved decreased greatly with volume. One consultant has estimated that:

> The cost of handling a container drops from $100 to approximately $30 as the weekly volume increases from 500 containers to 1500 containers. Although the capacity of a typical container berth is at least 1500 containers per week, many berths are currently used for only one ship per week. As a result, the cost to steamship lines is close to $100 per container.[10]

If each ship handled 2000 containers (1000 unloaded, 1000 loaded) per twenty-one-day round trip voyage, and made seventeen voyages per year, then each vessel handled 34,000 containers per year.[d] In order to be more conservative than the consultant's estimate, it was assumed that the cost of handling a container was $60 for the larger company and $100 for the smaller company. Consequently, the larger firm saved $1.36 million per ship per year on handling costs.[e] If this comparison had been between a large unsubsidized operator and a small subsidized operator, the savings would have more than eliminated the advantage of an annual operating differential subsidy appropriation of $700,000 per ship.

The Number of Containers Required in the System

One method of estimating the number of containers required used the following equation derived by the National Academy of Sciences for a multiship-multiport system:[f]

$$N = (n + i)C$$

where N was the maximum number of containers required, n was the number of ports served; i was the number of ships in the fleet; and C was the number of containers carried on each ship. Using this equation the small firm needed 3500 containers, while the large company required 19,500.[g] Hence with

---

[d] One week was lost for maintenance each year.

[e] Each ship handled 34,000 containers x $40 saved per container = $1.36 million. This analysis does not consider additional costs which the larger operator might have had to pay for additional terminal space or equipment to handle the larger volume of containers.

[f] This equation was derived in NAS-NRC, *Maritime Transportation of Unitized Cargo,* Publ.

[g] For the small firm, $N = (4 + 3)\ 500 = 3500$. For the large firm, $N = (12 + 27)\ 500 = 19{,}500$.

9 times as many ships, the large company only needed 5.6 times as many containers. Converting these figures to show the number of containers needed per ship, indicates that the large company required 445 fewer containers per vessel.[h] Containership operators were also required to buy or lease bogies which contained the wheels and suspension systems, etc., to use when moving their containers inland. If the number of bogies purchased was one third the number of containers, the large firm saved 148 bogies per ship in its fleet. The costs of containers and bogies varied widely. Assume that the cost of a forty-foot container was $3500 and the cost of a bogie was $4500. If three sets of containers and bogies were needed during the twenty-five year life of a ship, the total cost savings on an annual basis equaled $267,000 per ship.[i]

Operation of a Computerized Container Control System

Economies of scale also existed in the computerized container control system. Once the system was in operation, additional containers were added at low incremental cost. The cost savings here were less than in the other two examples so that no attempt is made to pinpoint this value.

Implications for Planning

This crude illustration has shown that definite economies of scale exist for large firms carrying unitized cargo. In theory, a small firm could gain these benefits by joining together with other firms in terminal facilities and container pools. In practice, however, companies sharing terminal facilities have sometimes complained of poor service and little control over their operations. No large-scale container pool has yet been formed, evidently because companies do not wish to have others controlling their equipment. Consequently, a key factor in planning must be determining how a subsidized line can best achieve the advantages of economies of scale in unitized cargo operations. Another factor with which the subsidized companies must contend is the new methods used to allocate CDS funds.

Allocation of CDS Funds

CDS funds are appropriated by Congress and allocated by the Maritime Subsidy Board. During a Congressional hearing in 1966 Peter Teige, an executive of American President Lines reported that "by far the most important

[h]Each vessel of the small company required 1167 containers, while each vessel of the large company required 722 containers. The difference was 445 containers.

[i]The total equipment saving over twenty-five years equaled $6,670,500 per ship. Divided by twenty-five gave $267,000 per ship per year.

and in fact the overriding reason for delay has been the failure of the government to appropriate sufficient funds for shipyard subsidies."[11] At the time of the hearing, American President Lines had already waited two years for CDS funds and was expecting a further postponement. Since obtaining such funds is critical to receiving new construction, a subsidized line must consider the Maritime Subsidy Board's allocations as part of its planning process. Contracts for new subsidized ship construction are listed in Table 4-4 for fiscal years 1958-1969.

Table 4-4
New Construction Contracts for Subsidized Combination and Cargo Ships Fiscal Years 1958-1969*

	Light Ship Long Tons	*Number of Ships*	*Design*	*Owner*	*Commercial Bid for Ship (Thousands of Dollars)*	*Foreign Cost (Operator's cost) (Thousands of Dollars)*
F. Y. 1958						
	6,600	2	C3-S-38a	AEIL	11,270	5,850
	6,600	2	C3-S-38a	AEIL	11,692	5,850
	5,870	5	C3-S-37a	Lykes	9,610	5,330
	5,963	2	C3-S-33a	M-M	10,491	5,590
	5,963	2	C3-S-33a	M-M	10,937	5,590
	8,550	2	C4-S-1q	APL	14,406	7,275
F. Y. 1959						
	5,870	4	C3-S-37a	Lykes	9,116	5,050
	5,963	3	C3-S-33a	M-M	9,985	5,250
	6,700	4	C3-S-46a	AEIL	10,899	5,555
	6,220	3	C3-S-43a	Miss.	9,591	4,990
F Y. 1960						
	7,870	3	C4-S-1s	AML	11,925	6,150
	8,600	2	C4-S-lt	PFEL	13,566	6,850
	10,600	3	C4-S1-49a	Grace	17,495	8,450
	8,250	4	C4-S-1u	States	11,071	6,030
	5,963	1	C3-S-33a	M-M	9,383	4,580
F. Y. 1961						
	5,910	4	C3-S-37b	Lykes	9,006	4,229
	7,600	6	C4-S-58a	Farrell	10,269	5,289
	7,850	5	C4-S-57a	USL	10,495	5,146
	6,700	4	C3-S-46b	AEIL	8,713	4,525
	7,850	6	C4-S-57a	USL	9,703	5,150
	5,980	4	C3-S-37c	Lykes	8,485	4,170
	8,250	2	C4-S-lu	States	11,039	5,770
F. Y. 1962						
	5,980	4	C3-S-37c	Lykes	8,018	4,049
	6,003	2	C3-S-37d	G & SA	8,145	4,201
	10,600	1	C4-S1-49a	Grace	17,907	8,896
	7,800	6	C4-S-60a	M-M	10,362	5,266
F. Y. 1963						
	7,533	5	C4-S-64a	USL	10,572	5,430
	7,848	2	C4-S-1sa	AML	11,278	5,344
	7,345	4	C4-S-66a	Lykes	10,039	5,037
	9,118	4	C4-S-65a	Grace	13,145	6,034
	8,500	3	C4-S-1qa	APL	12,378	5,623

Table 4-4 (cont.)
New Construction Contracts for Subsidized Combination and Cargo Ships Fiscal Years 1958–1969

	Light Ship Long Tons	*Number of Ships*	*Design*	*Owner*	*Commercial Bid for Ship (Thousands of Dollars)*	*Foreign Cost (Operator's cost) (Thousands of Dollars)*
F. Y. 1964						
	6,003	3	C3-S-37d	G & SA	8,560	3,944
	7,345	4	C4-S-66a	Lykes	10,328	4,642
	9,118	2	C4-S-65a	Grace	12,620	5,984
	7,132	2	C4-S-64b	Prud.	10,731	4,992
	7,345	4	C4-S-66a	Lykes	10,499	4,832
F. Y. 1965						
	8,064	5	C4-S-68a†	USL	12,190	5,743
	8,400	4	C4-S-69a	APL	13,706	6,302
	5,850	5	C3-S-76a	Delta	10,429	4,870
F. Y. 1966						
	8,400	1	C4-S-69a	APL	13,180	6,247
	8,480	3	C4-S-73a†	AEIL	12,923	5,913
	8,350	5	C4-S-69b	States	14,540	6,540
	9,701	3	C5-S-75a	AML	16,290	7,106
	9,810	1	C5-S-75b	PFEL	16,799	7,196
	11,139	4	C5-S-78a	M-M	15,698	7,900
F. Y. 1967						
	9,992	1	C5-S-75a	AML	15,670	7,047
F. Y. 1968						
	13,800	6	C8-S-81b	PFEL	21,313	10,614
	13,800	5	C8-S-81b	Prud.	21,313	10,628
	13,850	1	C7-S-68e	USL	15,750	7,822
F. Y. 1969						
	10,240	3	C6-8-85b	APL	20,910	10,160
	10,675	4	C6-8-85b	Farrell	20,950	9,812
	17,317	3	C8-S-82a	Lykes	32,474	14,613

*The cost figures in this exhibit do not necessarily include all contract changes which occurred after the initial contract was awarded. The following abbreviations are used in the exhibit:

AEIL - American Export Isbrandtsen Lines, AML - American Mail Line, APL - American President Lines, G & SA - Gulf and South America, Miss. - Mississippi Shipping Co., M-M - Moore-McCormack Lines, PFEL - Pacific Far East Lines, Prud. - Prudential Lines, USL - U.S. Lines.

†These designs were converted to larger full containerships while under construction.

Source: Maritime Administration.

Organization and Functions

The Maritime Subsidy Board is composed of three members: the Maritime Administrator, as chairman, the Deputy Maritime Administrator, and the General Counsel. The Secretary of the Board serves as an Alternate. The board performs the functions and exercises the authority vested in the Secretary of Commerce to award, amend, and terminate operating and construction differential subsidy contracts. It conducts hearings, makes determinations, and investigates the relative costs of building and operating ships in the United States and abroad. Decisions and orders of the board are final, unless within specified periods of time the Secretary of Commerce enters a written order stating that he elects to review the action of the board.

Situation During the Mid-Sixties

By the mid-sixties the Stabilized Shipbuilding Program was running behind schedule and many subsidized lines were anxious to construct new ships. A lack of funds appropriated for CDS purposes had been a frequent problem in the shipbuilding program.

In November 1965, the Maritime Subsidy Board issued the following statement:

> In making its determination (of CDS) under this policy, the Maritime Subsidy Board will take into consideration the following factors:
> 1. Number of vessels proposed for construction by the applicant.
> 2. Cubic and deadweight capacities and speed of the proposed vessels.
> 3. Proposed cargo handling equipment and techniques for transfer of cargo in and out of vessels and to and from inland points. In this connection the applicant will be required to set forth the estimated rate of loading and of discharge of cargo, as well as the adaptability of the proposed vessel to integrated systems of transportation embracing both ocean and overland transportation.
> 4. Estimated domestic cost of construction.
> 5. Estimated revenues and cost of operation; and with respect to wage cost, the proposed manning schedule on the proposed vessels.
> 6. The applicant's intention to seek operating subsidy and if so, the duration and the amount of such subsidy payments.
>
> The Board will weigh the above factors in such a fashion as will measure the productivity of the vessel (i.e., its carrying capacity, speed and rate of cargo handling) against the government's cost of construction and operating aid. The Board will award aid (so far as funds are available) for the construction of those vessels otherwise eligible, as will give the greatest productivity for each dollar of government aid.[12]

Consideration was also given to such factors as the age of the ships to be replaced, the trade route requirements, and the operator's financial situation.[13]

Request for Applicants

In April 1966, the Maritime Subsidy Board requested that subsidized lines having replacement obligations in 1967 and 1968 propose plans for joint shipbuilding providing for the same types of ships to be built in one shipyard. The idea of the board was to plan, for the first time, an entire fiscal year's construction subsidy program—and even beyond that year—with the end in mind of minimizing the different types of ships to be built, and hopefully achieving some price reduction through multiple-ship contracts.

The board stressed productivity in new ship designs. By productivity, the government meant the expected ton-miles produced per dollar of subsidy funds. A productivity formula was devised taking into account speed, capacity, sea time, utilization, construction-differential and operating-differential subsidy. Because the expected utilization or load factor was part of this formula, the proposed ships of the same design to be operated by two different companies on the same trade routes could have different productivity factors. The productivity factor bore no relationship to the operator's expected profit or return on investment.

Emphasis on productivity has caused the phasing out of construction of smaller ships. Construction contracts for subsidized C-3 vessels were last awarded in fiscal year 1966. The following year was the last in which contracts for C-4 vessels were awarded. It is doubtful that any more of these less productive types of vessels will receive CDS funds. In fact, C-5 vessels may also be excluded from further construction programs. A construction contract for a C-5 vessel was last awarded in fiscal year 1967. Since then all new ships have been larger and more productive.

In October 1966, the Maritime Subsidy Board presented a tentative ship construction program made up of the proposed ships with the highest productivity factors.[j] In making this decision, the Maritime Subsidy Board allocated CDS funds to build three ships of a new innovative design which would earn a return on investment of 0.6 percent according to the calculations of the Office of Government Aid.[k] Other subsidized lines with designs promising higher return on investments were bypassed to enable the construction of the more productive ships. The allocation of funds in fiscal years 1967–1969 essentially followed this proposed program with additional ships being ordered by some companies.

The point being made here is not that the board has acted capriciously. In the instance above, the construction of the three productive vessels will allow the replacement of fifteen vessels. Consequently, the board will save ODS on twelve ships per year. In addition, the innovative new design might prove to be more profitable than estimated. The important fact to consider is simply

---

[j]The only exception was the allocation of CDS funds for one ship which comprised a follow-on contract to vessels already under construction.

[k]This information is given in "Maritime Subsidy Board," a case study written by the author in 1970 at the Harvard Graduate School of Business Administration. In this case study, the companies trade routes, and designs have been disguided. It is interesting to note that the subsidized line proposing the 0.6 percent R. O. I. design had greatly underestimated its construction cost. After spending years developing this new design, the company apparently felt it had no alternative but to proceed with the construction.

that the Board has different objectives than the subsidized operator. While the company attempts to maximize its return on investment, the board has placed more emphasis on ton-miles produced per dollar of subsidy and possible savings in ODS.

Whether a subsidized line can improve its chances of obtaining funds by modifying its proposed vessel design is a problem without a simple solution. If a company modifies its design to make it more productive or joins another company in a multiship, multicompany, standardized design proposal, these actions would not hinder the company's efforts to obtain CDS funds. However, the issue is clouded by such other factors as the number of vessels to be replaced, the U.S.-flag market share on the trade route involved, the financial condition of the company, the personalities of the individuals involved, the length of time a firm has been waiting, the degree of innovation, and the number of vessels to be constructed. It would seem that informal negotiations with the Maritime Subsidy Board would be the best way to approach the problem of design modifications.

The clever actions of one subsidized line aided its obtaining CDS funds. On June 12, 1969, the Office of the Budget reported to the Maritime Subsidy Board that \$33,750,000 were still available for CDS funds in fiscal year 1969, which would end in twenty-eight days. Shortly thereafter the subsidized operator submitted a well-prepared application for \$33,650,000 in CDS funds for the conversion of eight vessels. This timely presentation resulted in the award of the funds.[14] In conclusion there is reason to believe that a company can improve its chances of acquiring CDS funds with proper planning.

Need for Better Planning

The combination of new factors in the environment and a high degree of risk involved in starting a unitized cargo operation places an increased need for long-range planning on the subsidized lines. The traditional vessel replacement planning process must be improved to make ship replacement an integral part of long-range planning. Figure 4-3 summarizes diagrammatically the unstable environment with unitization.

The research of others apparently had little effect on the subsidized lines. After Sea-Land began full containership service on the North Atlantic in April 1966, more than thirty subsidized break-bulk cargo vessels and semi-containerships were launched.[1] Three years later, many subsidized companies regretted their decisions. A survey presented at a congressional hearing in May 1969 showed that eight subsidized lines wanted to add full containerships to their fleets. Details are given in Table 4-5. Chapter 5 shows that poor planning accounted for many mistakes in the replacement of subsidized ships.

[1]In due respect to American Export Isbrandtsen Lines, it should be noted that this company began full containership service on the North Atlantic a few months after Sea-Land.

Door-To-Door Movement

Specialized Ship Design and Terminal Facilities

↓

Shipper Can Differentiate Among Competing Liner Services

↓

Unstable Market Share

↓

Very High Penalty Cost for Designing Non-Optimal Ship

↓

Increased Need For Long-Range Planning

↓

Need For Improving Traditional Ship Replacement Planning Process

Figure 4-3. Unstable Environment With Unitization.

Table 4-5
Summary of Liner Vessels Which U.S. Subsidized Companies Would Contract for During Fiscal Year 1970 if Sufficient CDS Funds Were Available

Company	*Type of Ships*	*No.*	*Estimated Total Cost (millions of dollars)*	*Estimated CDS at 55 Percent (millions of dollars)*
AEIL	Containerships	6	148	81.0
AML	Containerships	2	44	24.0
APL	Containerships	2	44	24.0
Delta	Containerships	3	60	33.0
Lykes	Conversions to Containerships	9	18	9.9
Oceanic	Containerships	2	40	22.0
Prudential*	LASH	5	118	64.9
States	Containerships	5	107.5	59.1
U.S. Lines	Conversions to Containerships	8	64	35.2
	Total	42	643.5	353.1

*Conditional on the approval of the acquisition of Grace Lines.

Source: U.S. House of Representatives, Committee on Merchant Marine and Fisheries, "Maritime Authorization - 1970," 91st Congress, 1st sess., May 9, 1969, p. 4.

5 Subsidized Ship Replacement Planning Process: 1961 and Beyond

Steamship owners in the ocean-going trade have always had a reputation for being conservative. McDowell wrote "tradition, precedent, and the time-tested practices of building and operation provide security and certainty. There is a natural reluctance to introduce change, for even the simplest modification of a bill of lading, of an accounting procedure, or of a structural characteristic of a ship may open up a variety of unprecedented situations with respect to legal interpretations and responsibilities."[1]

With a history of conservatism and an environment which created little need for long-range planning, the subsidized steamship companies were not famous for great efforts in the area of research and development. In 1965, the Maritime Administrator, Nicholas Johnson, stated,

> I will stand by my statement that there is, among American subsidized operators, still, no Vice President whose sole job it is to direct the research and development efforts of that company . . . . There are, of course, research and development efforts in many shipping companies in America today. My point is only that they are not as great as they might be.[2]

More recently, the subsidized industry has been the target of criticism of a more caustic nature. The American Maritime Association stated, "it is an open secret that the subsidized fleet in the early and middle sixties was a vast blunder, an era of business judgment whose monument is a generation of ships commercially obsolete before, in many instances, leaving the yard, and which no one pretends can compete against container ships."[3]

The AMA (American Maritime Association) is made up largely of unsubsidized lines that are often in bitter competition with the subsidized firms. However, even persons within the subsidized lines have also commented adversely on their industry. Joseph A. Medernach, Vice-President of Moore-McCormack Lines, stated that "the industry is one of the most backward, stodgiest, and stuffiest businesses around."[4] Admiral Will, the chairman of the board of American Export Isbrandtsen Lines (AEIL) wrote, "archaic business methods and the failure to exploit the fruits of technical developments contributed measurably to the myriad problems we face today."[5]

Methodology of the Analysis

To analyze recent ship replacement processes of subsidized lines, four shipping companies were studied from the time they began design work on their latest vessels until these ships were delivered. In each instance, these vessels were the first new construction used by each firm in the era of unitization. The planning processes used by each firm showed considerable room for improvement.

The information on each company is analyzed in the following format:

1. Initial Strategy

After giving some background on the company, the section describes the initial strategy explicitly articulated by the company or deduced from its past behavior as it began the design work for its most recent ship construction. Interviews with the management of the firms described here were conducted to verify these strategies.

2. Planning for the Initial Vessel Design

The original ship design chosen by the company is described in such terms as size, speed, and container capacity. The processes leading up to this design are reported to show how the initial strategy was followed.

3. Changes in the Environment and Comparison with Other Ship Designs or Changes in the Ship Design

Many years expired from the time each company began design work on a new vessel until the last of the new flight of ships was in service. Although as each firm started designing its new vessels, it was fully aware of the trend towards unitization, the company's initial evaluation of its environment did not necessarily remain the same during the period of vessel construction or after delivery of the new ships. This section describes how the actions of each firm's competitors changed the environment and how the subsidized lines studied reacted in terms of design changes.

4. Outcome of this Course of Events

The reactions of the companies to their final vessel designs were valuable in determining the final consequences of the planning processes. This outcome is described and assessed in this section. A subsidized line that was dissatisfied with some of the results of its planning process has paid some penalty for its mistakes: A company might have finally built a highly productive vessel but only after incurring unnecessary costs of funds and time delays; or on the other hand, a firm might have produced ships which were not economically competitive.

5. *Weakness in the Planning Process*

A company that was dissatisfied with the results of its planning process probably has weaknesses in its decisionmaking process. By analyzing how this dissatisfaction resulted from the initial strategy and planning process used, the weaknesses are determined in this section. A company may have failed to perform an analysis of its environment or may have simply conducted a superficial analysis. A firm might have performed an analysis which ignored or did not adequately consider a key element, such as possible cargo diversion or possible actions by the competition. A failure to understand fully the economics of unitization could have resulted in a company trying to follow a strategy which had poor financial consequences. Concentration on the immediate environment without adequate analysis of possible changes in the future could have also caused economical penalties to the company. Whatever the planning weaknesses, they are described in the sections that follow.

States Steamship Company

1. *Initial Strategy*

States Steamship Company operates on trade route 29 between the Far East and the U.S. West Coast with San Francisco its main domestic port of call. One of the smaller subsidized lines, the company had only thirteen ships in 1965. At that time, States felt that containerization might eventually come to its trade route. The firm realized that many obstacles would hinder it from starting a container operation. As a small company, States did not have the staff necessary to analyze cargo forecasts, design new vessels, study inland transportation networks, or establish a computerized container control system. The line was also reluctant to invest large sums of money to enter a style of operation in which it had no experience. Consequently, States decided to continue as a break-bulk cargo operator, feeling that even if large-scale containerization came to its trade route, its small size would permit it to remain profitable carrying break-bulk cargo.

2. *Planning for the Initial Vessel Design*

Having decided to continue as a break-bulk cargo operator, States looked closely at the new designs of its break-bulk cargo competitors. American President Lines, a U.S.-flag-subsidized competitor providing service between California and the Far East contracted for four C-4's in fiscal year 1965 and for an additional ship of identical design in September 1965. States Steamship decided to use almost exactly the same design and contracted for five ships in May 1966. The ships were scheduled to be delivered between late 1968 and the middle of 1969.

The States vessels cost the company approximately $6.94 million per ship. They have a bale capacity of 775,000 cubic feet, a deadweight tonnage of 14,250, and a service speed of 23 knots. These break-bulk cargo vessels can carry up to 216 twenty-foot containers, but they are not designed to carry forty-foot containers.

3. *Changes in the Environment and Comparison with Other Ship Designs*

By the time States had awarded the contract for its new ships, Sea-Land had already begun its commercial, full containership service on the North Atlantic as well as a full containership service for the Department of Defense between the U.S. West Coast and Okinawa. In addition, American President Lines had applied to the Maritime Subsidy Board for CDS to convert four C-4 vessels to full containerships. Consequently, States might have had good reason to believe that they could not continue indefinitely as a break-bulk operator.

In addition to choosing to remain as a break-bulk operator, States ignored American Mail Line's huge break-bulk cargo ship design and adopted the less productive C-4 design. AML's C-5 design, for which a contract was awarded in June 1966, cost the company about $7.67 million per ship. However, the AML vessels have a bale capacity of 1,054,130 cubic feet and a deadweight tonnage of 20,978. The service speed of the C-5s is 20.8 knots. Thus, the AML ships have about 36 percent more bale capacity and approximately 47 percent more deadweight tonnage than the States ships, while only costing about 11 percent more.

Although the States ships have a higher service speed than the C-5s, the C-4s requirement for higher horsepower (24,000 H.P. for the States' ships versus 21,590 H.P. for the C-5s) results in a higher fuel bill. There are advantages to the higher speed in the C-4 vessels, although the benefits are difficult to quantify. The higher speed might provide a marketing advantage for States Steamship and might give it more flexibility in route schedules.

In early 1967, the Japanese government which had previously discouraged large-scale containerization, started planning low-interest, containership building loans and containerport imporvements. By the end of 1967, Matson Navigation, a U.S.-flag unsubsidized line, had begun a full containership service between California and the Far East, and other competitors had either made plans to build full containerships or already had them under construction.

In June 1968, a study by Litton Systems stated that 89.6 percent of the exports and 96.6 percent of the imports on trade route 29 were containerizable.[6] A few months later a top executive of States maintained that his company could still continue as a profitable break-bulk cargo operator. States Steamship had no desire to disrupt its organization by starting container operations. Management felt that the ships under construction could still show a profit while carrying mostly break-bulk cargo. Consequently, the container capacity of the new ships was not raised above 216.

In a discussion with the author, this executive stated that if more funds were allocated for planning, he would not spend it on market research. First, he did not have a staff to do the work. Second, he did not feel that it was worthwhile to pay outside consultants to do the job. He remarked that his last contract with an outside consultant for a modest sum had been an unfortunate experience. Although the company had spent little time analyzing its environment, experience and intuition led this executive to believe that the profitability of States break-bulk cargo operations would continue in the future.

In February 1969, several months before States Steamship received the last of its new flight of ships, United States Lines, a large subsidized line operating from the U.S. Atlantic Coast announced the sale of five modern C-4 type vessels at a price of approximately $4.5 million each. The ships, which were only four or five years old, had a deadweight tonnage of 13,264 long tons, a bale capacity of 687,057 cubic feet, a service speed of 20.8 knots, and container capacity for 182 twenty-foot equivalent containers. United States Lines had apparently decided that with the arrival of full containerships modern break-bulk cargo vessels were no longer competitive.

Compared to the United States Lines ships, States' new vessels had about 13 percent greater bale capacity, approximately 7 percent greater deadweight tonnage, and about 45 percent greater cost. States Steamship would receive ships that were four to five years younger with a 2.2 knot faster service speed and a container capacity that was thirty-four containers greater than the vessels U.S. Lines had abandoned. Since the new States's vessels would be operating on the highly competitive trade route between the U.S. West Coast and the Far East against full containerships, it seemed that these ships, too, were obsolete.

4. *Outcome of this Course of Events*

States changed its strategy in 1969. A survey presented in a congressional hearing in May 1969 showed that States would contract for five full containerships in fiscal year 1970 at a cost to the company of almost $10 million per ship if CDS funds were available. Later States decided to build roll-on/roll-off vessels rather than full containerships.

States' decision to start a large-scale container operation came only after full containerships had been serving its trade route for about two years. When States finally puts roll-on/roll-off vessels in service, its competitors will have their container operations firmly established. Since States had been very reluctant to change its style of operations until recently, it seems probable that its investment of almost $35 million in new break-bulk cargo vessels will not be profitable in the future.

5. *Weakness in the Planning Process*

States has essentially avoided the problem of long-range planning. Although experience and intuition were probably valuable aids in the stable break-bulk cargo environment, the trend to unitization has made planning and analysis necessary. An examination of cargo flow on States' trade route would have shown the company what types of cargo would not be carried on full containerships. An analysis of the relevant factors could have been performed in 1965 while the firm was considering new construction. By continuously updating this analysis, States would have arrived at its present discovery of the necessity of carrying unitized cargo long before 1969. Consequently, States could have converted its new vessels into roll-on/roll-off vessels or full containerships while the ships were still under construction.

Because States is planning to build only a small number of roll-on/roll-off vessels, the firm will have a difficult time taking advantage of the economies of scale with unitization. If a long-range planning staff had been established, this group would have had the opportunity to evaluate such alternative strategies as selling the new vessels while under construction to one of States' competitors who would have been eager to convert them to full containerships, getting out of the steamship business entirely, going off of subsidy and becoming a tramp operator, merging with another firm, or forming a consortium. By failing to plan ahead, States had already missed a valuable opportunity to improve its fleet and its future financial position.

American Mail Line[a]

1. *Initial Strategy*

In 1964, American Mail Line was one of the smaller subsidized lines with only nine vessels. Operations between the U.S. Pacific Northwest and

[a]For More information on this company, read "American Mail Line Ltd.", a case study written by the author in 1969 at the Harvard Graduate School of Business Administration.

the Far East on trade route 29 provided most of the firm's revenues and profits. AML also served Indonesia, Malaysia, and India on trade route 17 from the U.S. Pacific Northwest.

Toward the middle sixties, American Mail Line performed an analysis of its environment while planning for new vessel construction for trade route 29. The study focused on the cargo flow data of the company and the steamship conferences to which it belonged. AML discovered that more than 90 percent of the imports from the Far East to the U.S. were containerizable. However, almost none of the commercial exports leaving from the U.S. Pacific Northwest to the Far East were containerizable. Consequently, AML decided that with such as imbalance of containerizable cargo full containerships would not be used on its trade route. Therefore, the company decided to continue its current style of operations with the building of break-bulk cargo ships.

2. *Planning for the Initial Vessel Design*

Since AML had been quite successful with the use of C-4, Mariner design ships, it favored replacing the four warbuilt vessels in its fleet with ships of a modified Mariner design. The Maritime Subsidy Board, responsible for allocating CDS funds, was emphasizing productivity in new vessel designs and persuaded the company to build C-5 design vessels. Although these huge ships were 605 feet long with a deadweight tonnage of 21,590 tons, they could only carry 411 twenty-foot equivalent containers since the design was based on a break-bulk cargo operation. Consequently, when the C-5 design was fully loaded with containers, only approximately 50 percent of its bale cubic capacity was utilized.

3. *Changes in the Environment and Comparison with Other Ship Designs*

A contract was awarded in June 1966 for construction of four of these ships, three to be owned by American Mail Lines and one to be delivered to Pacific Far East Lines. Later, American Mail Lines bought this ship from the other subsidized operator. Earlier in 1966, American President Lines had applied to the Maritime Administration for CDS funds to convert four C-4 Mariner-type ships to full containerships for express California–Far East voyages. The vast majority of the commercial liner cargo going to the Far East from ports in California was containerizable. Consequently the service between California and the Far East would eventually be dominated by full containerships.

One month after the initial AML construction contract had been awarded, Sea-Land began a full containership for the Defense Department between the U.S. Pacific Coast and Okinawa. This service demonstrated that 40,000 measurement tons of containerized cargo could be off-loaded and loaded in 750 man-hours, a job that would have taken 24,000 man-hours with break-bulk cargo in conventional ships. By this time, Sea-Land had already been providing a full containership service on the North Atlantic for more than a year, and competing operators had full containerships in service or under construction. In the Far East, the Japanese government had helped to discourage large-scale containerization for many years; but in early 1967, it was planning container port improvements and low-interest containership building loans to its shipping lines.

Although AML realized that full containerships could greatly reduce costs on trade routes possessing large amounts of containerizable cargo, it still felt that the great imbalance in containerizable cargo between the Pacific Northwest and the Far East would make the use of full containerships uneconomical between these regions.

In the summer of 1968, the management of AML discovered that three of its new ships could possibly earn a greater profit carrying nothing but containers on a shuttle run between Japan and the U.S. Pacific Northwest, rather than operating as break-bulk cargo carriers, as they were designed.[b] The company felt that the cargo loadings used in its calculations were realistic estimates of its prospective business with the delivery of the new ships. With the new ships running totally as break-bulk cargo carriers at 71.5 percent load factor, each ship would earn a net profit after taxes of $2,320 per day. If three of these vessels ran as container shuttles with just 411 twenty-foot equivalent container spaces each (a 62.2 percent load factor), each ship would produce a net profit after taxes of $2,060 per day. Eastbound each ship would carry 411 containers of commercial cargo. Westbound each ship would carry 100 containers of military cargo.

If American Mail Lines were able to attract twenty-two additional containers per round trip voyage, the net profit after taxes per day of the container operation would be equivalent to that of the break-bulk cargo operation. This would bring the load factor of the containership to 64.8 percent. Since the container system has a higher fixed cost and lower variable cost than the break-bulk cargo system, as cargo increased the incremental profit to the container system would be higher than with the break-bulk cargo system.

At the time AML was performing these calculations, outside consultants informed the company that large amounts of commercial containerizable cargo moving from the U.S. Midwest to Japan via San Francisco might be diverted through the port of Seattle. AML also learned that the Japanese planned to place full containerships with a capacity for 750 twenty-foot equivalent containers on the run between Japan and the U.S. Pacific Northwest in 1969. Nonetheless, AML decided not to change the design of the ships under construction. Each ship cost the Company $7,670,000.

If AML had converted the last three ships under construction to full containerships for the shuttle run to Japan, each vessel would have been

[b]The supporting data appears in "American Mail Line, Ltd."

able to carry about 1000 twenty-foot equivalent containers. American Export Isbrandtsen Lines, who had contracted to build three C-4 design conventional cargo ships in November 1965, later converted their ships to full containerships with a carrying capacity of 928 twenty-foot equivalent containers. Although they waived CDS on the additional $500,000 cost per ship for conversion, the total cost to the company for each ship was only about $6,630,000. The larger C-5 vessels of AML could have been converted to have greater carrying capacity than the AEIL containerships.

4. *Outcome of this Course of Events*

In 1969, the competition put full containerships on AML's major trade route, This apparently impressed upon AML the economic benefits of full containerships, because a survey presented in a congressional hearing in May 1969 showed that AML would contract for two full containerships in fiscal year 1970 if CDS funds were available. These ships would cost the subsidized line approximately $10 million each. Later in the same year, AML decided to convert some of its C-4 ships to full containerships, jumboizing them in the process. These conversions would apparently take the place of any further new construction and cost approximately $4 million per ship to the company.

Since AML did not decide to obtain full containerships until 1969, the company did not place these vessels in service until after its competitors' container operations were well established. AML's decision to acquire full containerships apparently resulted either from a discovery that the new C-5 vessels were having difficulty competing against full containerships or an expectation of such difficulty in the future. Consequently, the company's $38 million investment in C-5 vessels may have difficulty showing a profit.

5. *Weakness in the Planning Process*

Although AML faces an outcome similar to States, the factors behind this result are quite different. Unlike States, AML confronted the problem of long-range planning in the initial stages of its vessel replacement program. Although AML had performed an analysis of its environment, two main weaknesses existed in its planning process: (1) the failure of the company to evaluate thoroughly the potential of inland transportation for increasing the amount of containerizable cargo in its market, and (2) the underestimation of the economic consequences of competing against full containerships. In retrospect, AML also underestimated the ability to place bulk-type cargoes, such as paper pulp, into containers to boost the load factor in the westbound or outbound direction. If a more intensive examination had been made of these factors, AML would have reached its decision on full containerships in time to convert its vessels under construction. Since AML is planning a small containership fleet, the company faces the same difficulties as States in taking advantage of economies of scale. As with States, further emphasis on long-range planning might have led to the analysis of such alternatives as forming a consortium, going off subsidy and becoming a tramp operator, selling the C-5s while under construction to a company eager to convert them to full containerships, or going out of the steamship business.

Moore-McCormack Lines[c]

1. *Initial Strategy*

When the Ship Replacement Program began in 1956, Moore-McCormack Lines agreed to replace forty-one ships over a period of several years. By the middle sixties, fourteen new ships had either been delivered or were under construction for the company. At that time, Moore-McCormack was one of the larger subsidized companies and operated the following three cargo services: (1) the American Republics Line which served trade route 1 between the U.S. Atlantic Coast and the East Coast of South America, (2) the Robin Line which operated on trade route 15A between the U.S. Atlantic Coast and South and East Africa, and (3) the American Scantic Line which operated on trade routes 6 and 8 between the U.S. North Atlantic and Scandanavia, the Baltic ports, Belgium, and the Netherlands.

In 1965, when Moore-McCormack was starting design work on its next flight of ships, several companies were considering starting full containership service on the North Atlantic, but there was not yet a large scale movement of containers in the U.S. foreign trade. With this uncertainty, Mooremac desired great flexibility in its operations. It felt the need for a design suitable for all the company's services, and also suitable for charter to the military when necessary. With such a very flexible ship design, Moore-McCormack apparently felt that it could react to any change in the environment.

2. *Planning for the Initial Vessel Design*

The C-5 design Moore-McCormack created was unique. The ship had 30-ton cargo handling booms so that it could handle break-bulk cargo in the conventional manner. It also had a 70-ton boom which the company

[c]For more information on this company read "Moore-McCormack Lines, Inc., Analysis of Roll-On/Roll-Off Ships," a case study written by the author in 1969 for the Harvard Graduate School of Business Administration.

did not want but came with the ship as a National Defense Feature. In addition, when the hatches were closed on the next to top deck, vehicles could be rolled aboard by means of a stern ramp. When the hatches were opened, slides on the bottom of the hatches would act as cellular holds for the ship. The ship could carry 830 twenty-foot equivalent containers when serving as a full containership. For maximum usage of the roll-on/roll-off deck, a false deck could be rigged to permit the storage of 484 foreign cars. At the same time, less than 600 twenty-foot equivalent containers could be carried. Commenting on this unusual C-5 design, the *Journal of Commerce* had this opinion: "Though such a mixture of capabilities in one ship sounds slightly incongruous, it reflects Mooremac's decision not to try to guess the trend of events but be ready for whatever happens."[7]

The roll-on/roll-off (ro-ro) feature of the four new ships would give Moore-McCormack the only U.S.-flag vessels in the U.S. commercial foreign trade with this characteristic. However, U.S.-flag competitors carried automobiles on deck or in containers. A foreign consortium, Atlantic Container Line, had under construction four vessels with a ro-ro capacity more than double that of the Moore-McCormack ships. Shortly after the ACL vessels entered service in 1967, the consortium ordered six larger ships for their fleet. Again the second generation ships were to have more than twice the ro-ro capacity of Moore-McCormack vessels. In addition, the Volkswagon Corporation, which exports the most cars to the U.S., provided oceanborne transportation for its automobiles through a fleet it controlled. The Journal of Commerce remarked of the Mooremack ro-ro feature, "While there may be some commercial advantage from such an operation it is doubted that Mooremac relies very heavily upon it."[8]

Moore-McCormack's manager of new construction wrote, "At the time of the original design, the Military Sea Transportation Service, through the medium of an RFP, set forth certain requirements for the charter of the roll-on/roll-off–type vessel. This possibly influenced our thinking and led us to the design we now have."[9] However, subsidized lines with competitive trade routes and with ships of various ages, were normally reluctant to charter their best modern ships to the military if the ships could be effective on their commercial trade routes.

3. Changes in the Environment and Comparison with Other Ship Designs

By the end of 1965, several companies had been making plans to provide container service on the North Atlantic and at least two companies planned to use full containerships. In February 1966, Moore-McCormack started the first regularly scheduled container service on the North Atlantic by using up to 130 twenty-foot equivalent containers as deck cargo on its modern, conventional break-bulk, C-4 type vessels. These ships were later adapted to increase their capacity to 270 twenty-foot equivalents. Moore-McCormack obtained its containers from leasing companies.

In June, 1966, a contract was awarded to build four 25-knot ships of the unique C-5 design for Moore-McCormack. At that time the company planned that these new ships would provide a weekly service on the North Atlantic, on delivery in 1969, allowing the C-4s to replace older vessels on another trade route.

By carrying containers on the decks of its modern C-4 break-bulk freighters, Moore-McCormack increased the flexibility of the company's operations. However, Equity Research Associates, an investment research company, commenting on Moore-McCormack's use of C-4 semicontainerships wrote, "Moore-McCormack elected to enter containerization cautiously, and that option has proved expensive. . . . Combining container and break-bulk service on the same ship produces two easily predictable results; costs rise and ship efficiency drops. The operator adds the costs of leasing, handling, storing and positioning inland the containers, ship loading and unloading time is increased."[10]

The investment research firm gave the numbers used in Table 5-1 as an illustration of how the operation of a combination container/break-bulk ship compared to that of either a full containership or a break-bulk ship. The calculations were originally performed in 1966 by a steamship line serving the North Atlantic.

In 1966, Sea-Land and American Export Isbrandtsen Lines entered the North Atlantic with full containerships. In addition, U.S. Lines and AEIL began to convert ships under construction to full containerships to be used on the North Atlantic, removing all deck cranes in the process. The head of vessel replacement for U.S. Lines stated that using container cranes on land eliminated the need for shipboard equipment that would reduce by as much as 20 percent the carrying capacity of the full containership.[11]

In May 1967, a study by the Port of New York Authority estimated that approximately 91 percent of the imports and about 86 percent of the exports on the North Atlantic trade routes 5, 7, 8, and 9 were containerizable.[12] By the middle of 1968, the Maritime Subsidy Board gave Moore-McCormack permission to expand its North Atlantic service to include trade routes 5, 7, and 9. Consequently, ports in the United Kingdom, Ireland, Germany, and in the Atlantic portion of France and Spain were now open to Moore-McCormack. If the company had taken the deck cranes off its new vessels, it would have had no difficulty finding containerports with shoreside cranes in its North Atlantic services.

In an interview with the author an executive of Moore-McCormack estimated that the combination of deck cranes and ro-ro equipment accounted for the loss of about 150 twenty-foot equivalent container spaces. The

Table 5-1
Economic Comparison of Full Containerships, Break-Bulk Cargo Vessels, and Semicontainerships

	Break-Bulk	*Semicontainer*	*Full Container*
Price of ship	$2,436,000	$5,880,000	$7,368,000
Cost of containers		504,000	3,528,000
Per Voyage			
Vessel expense	31,178	49,500	56,460
Cargo handling expense	124,408	110,880	131,160
Depreciation and overload	51,037	96,360	122,040
Total cost	206,623	256,740	309,660
Revenue earned	259,968	260,002	519,936
Pretax profit			
Per voyage	53,345	3,262	210,276
Per year	458,400	33,600	3,554,400
Ratio of annual profit to capital cost	18.8%	0.5%	32.6%

Source: Equity Research Associates, "Containerization," New York, May 16, 1969, p. 17.

deck cranes accounted for the vast majority of the lost container space. The final cost to the company would be about $8.15 million per ship, higher than a previous estimate of $7.9 million. The unusual features on the ship would cost the firm approximately $1 million per ship.

If the new Moore-McCormack vessels had been designed as full containerships, their container capacity would have been about 18 percent greater at a cost approximately 12 percent lower. If the new ships had kept the roll-on/roll-off features but eliminated the deck cranes, the increase in container capacity would have been almost as great, and the cost of the deck cranes would have been saved.

4. *Outcome of This Course of Events*

While Moore-McCormack was receiving delivery of its four new vessels in 1969 and 1970, Equity Research Associates predicted that Moore-McCormack would not be serving the North Atlantic by the end of 1970.[13] The investment research firm felt that Moore-McCormack's use of the flexible conbination container/break-bulk cargo C-4 ships had contributed heavily to the line's loss before federal taxes of $3.55 million in 1967 and $7.78 million in 1968.

Moore-McCormack's new ro-ro vessels met severe competition on the North Atlantic from full containerships and semicontainerships. The following statement to the stockholders from the 1970 annual report implied that Moore-McCormack regretted that it had not built full containships:

> As previously reported in the press, we teminated operations in the North Atlantic trade, in which this corporation had been engaged almost since its founding in 1913, and sold our four roll-on/roll-off container vessels to American Export Isbrandtsen Lines . . . . The conversion of the ro-ro vessels, which were designed as combination roll-on/roll-off container ships to meet then existing port facilities to full containerization would not have been justifiable in view of the existing and prospective conditions in the trade, and we reluctantly concluded that withdrawal from service was the only practical course of action open to us.

5. Weakness in the Planning Process

Moore-McCormack had avoided developing a corporate strategy for its long-range planning. Instead, the company had chosen to maintain a large degree of flexibility in general operations and in its new vessel design. Only after its new ships had been delivered did Moore-McCormack fully realize the disadvantages of having great flexibility in its new vessels. This initial design might have been avoided if proper planning had been performed to reveal the high costs of flexibility in a specialized market.

Executive Lines[d]

1. Initial Strategy

In the early sixties, Executive Lines was one of the largest subsidized steamship companies with more than forty ships in service or under construction. This subsidized line operated between the U.S. North Atlantic and the United Kingdom, Ireland, Germany, Belgium, the Netherlands, and the Atlantic portions of France and Spain on trade routes 5,7,8, and 9. Executive Lines also served the same general foreign area from the U.S. South Atlantic on trade route 11. In addition, the company provided service on trade route 12 between the U.S. Atlantic Coast and the Far East.

In 1963, Executive Lines was considering new construction to replace vessels on its North Atlantic trade routes. By this time the economic and technological feasibility of containerization for certain trade routes had been proved. However, the only attempt at using full containerships in the U.S. foreign trade had resulted in a dismal failure for Grace Lines in 1960 when Venezuelan longshoremen had refused to handle the containers. Executive Lines decided to design a break-bulk cargo ship that could be used for a gradual transition to containerization.

---

[d]The name of this company has been disguised. For more information on this firm, read "Executive Lines (A)," a case study written by the author in 1970 for The Harvard Graduate School of Business Administration.

2. *Planning for the Initial Design*

In 1964, Executive Lines completed work on a C-4 ship design that could carry 228 twenty-foot equivalent containers and had a service speed of 23 knots. This design resulted from the company's "policy of compromise and concession."[e] The low container capacity was a compromise between an ordinary break-bulk cargo vessel and the economic potential of a full containership. The design also represented a concession to longshoremen who were opposing the trend to large-scale unitization. Planning for the long-range future was secondary in Executive Line's view to withstanding the transition period to containerization.

3. *Changes in the Environment and the Ship Design*

A contract for five ships of this type was awarded in December 1964. The first ship was to be delivered in 810 days. In May 1966, one month after Sea-Land had successfully introduced full containership service on the North Atlantic, Executive Lines applied to the Maritime Administration to convert three of their ships under construction to medium-size full containerships, each carrying 638 twenty-foot equivalent containers. This change in design came about when " planning emphasis began to turn away from simple servicing of trade routes and ports toward new approaches for survival." Hence, once Sea-Land had surmounted any obstacles to using full containerships, Executive Lines' second ship design was affected by economic necessity rather than concern for labor or other parties.

These conversions required changing completely the design of the ships under construction. To obtain permission for this work from the Maritime Subsidy Board, Executive Lines offered to waive Construction Differential Subsidy on the added $2,250,000 cost per ship. At a CDS rate of 50 percent, Executive Lines was foregoing $3,375,000 in subsidy for the three ships.

In May 1967, Executive Lines applied to the Maritime Administration for a second design change. They obtained permission as well as CDS funds to jumbo-ize all five vessels to produce large containerships capable of carrying 1178 twenty-foot equivalents. The original power plants were maintained, putting a constraint on the service speed of about 22 knots.

About the time of this application, the Port of New York Authority published a study showing that 91 percent of the imports and 86 percent of the exports on Executive Lines' North Atlantic trade routes were containerizable.[14] Prior to submitting their application, Executive Lines had hired outside consultants who had convinced Executive Lines' management of the economies of scale possible with large containerships. In addition, by this time Sea-Land had made its presence felt on the North Atlantic. At a Congressional hearing in November 1967, Michael McEvoy, President of Sea-Land Service, Inc.,

[e]Quotations describing the attitude of the company are taken from the case study "Executive Lines (A)."

remarked regarding the company's North Atlantic service, "In the last 12 months alone, Sea-Land has collected freight revenues on that service of over $23 million. Our weekly service on this route is moving at capacity, and we have plans to add additional tonnage in 1968."[15] When Executive Lines felt the impact of Sea-Land's competition and discovered that "unless a full containership could realize economies of scale, it could not last long," it responded with the third design. This final ship design was influenced by the return on investment in the long run.

4. *Outcome of This Course of Events*

In May 1968, three and one-half years after the contract was awarded for the initial ship design, Executive Lines finally accepted delivery of the first ship of the latest design. This was the first U.S.-flag vessel ever built from the keel up as a full containership and also the fastest full containership in the U.S. Merchant Marine. After all its changes the company could boast of very productive new ships, but the firm might have avoided the additional cost of more than $3 million and a time delay of more than one year with better planning.

5. *Weakness in the Planning Process*

The basic weakness in Executive Lines' planning was its failure to use the data that were available or could have been developed for looking at the long-range picture. Executive Lines' first design, capable of carrying 228 containers, was based on the belief that a gradual transition to containerization would occur on the North Atlantic. However, well before Executive Lines had contracted for these vessels the National Academy of Sciences had performed studies showing that full containerships were economically far superior to semicontainerships.[16] Consequently, if Executive Lines' planning process had included careful analysis of this literature or had applied the cost methodology used in these reports to examine the feasibility of full containerships in its environment, the company might have designed a break-bulk cargo vessel which could have been easily converted into a full containership. Instead the conversion to a medium-size full containership required a complete redesign with a cost of $2,250,000.

Executive Lines decided on this conversion after Sea-Land had proven the feasibility of using full containerships on the North Atlantic. In may 1967, they decided to convert their vessels to large full containerships. However, the basic knowledge on which this decision was based could have been derived at the time of the earlier conversion in May 1966. Information on the amount of containerizable cargo and the economies of scale with larger containerships could have been obtained in the early and middle sixties. Consequently, the changes in vessel design in May 1966 could have included all the necessary alterations later devised by Executive Lines.

6 Environment and Fleet Improvement Planning Processes of Sea-Land Service and Matson Navigation Company

Protection of the Domestic Trades

Sea-Land and Matson were the first two companies to begin large-scale container operations. Both firms started experimenting with containers in the domestic trades in the late fifties. Vessels operating in the U.S. domestic trades were subject to cabotage laws which prohibited the use of foreign-flag vessels. The following paragraph from Professor Gorter's book summarized well the status of cabotage:

> Almost from its beginning as a sovereign nation the United States has restricted its intercoastal and coastwise trades to American-built, American-flag vessels. A law passed and approved in 1798 provided for discriminatory tonnage duties to be placed on foreign-operated vessels plying the coasting trade of the United States. Not only were the duties to be paid by foreigners higher but they had to be paid upon arrival at any United States port instead of once a year as required of United States-flag operators. In 1808, legislation was approved excluding foreign-flag vessels from the domestic trades. In 1817, another law was passed again excluding foreign flag vessels, from these trades. The Merchant Marine Act, 1920, applied similar restrictions to noncontiguous trades.[1]

The Merchant Marine Act of 1936 excluded subsidized lines from generally operating in the domestic trade. Consequently, Matson and Sea-Land were assured of protection from both foreign-flag lines and the U.S. subsidized lines as they conducted their operations in the domestic trade.

In the coastal and intercoastal trades, Matson and Sea-Land had to compete against railroads and truck lines. However, in noncontiguous routes such as Hawaii and Puerto Rico, they had, of course, no fear of competition from land transportation. The protection of the domestic trades was an important factor in bringing innovation to this area.

Dilemma of Operators in the Domestic Trades after World War II

After World War II, the operators in the U.S. domestic trades found themselves in a predicament. Labor costs had risen more rapidly in shipping operations than in railroading and trucking. This had been primarily responsible for placing the domestic ship operators in a profit squeeze between costs and revenues despite large increases in freight rates. Crews' wages, benefits, and subsistence had increased by 357 percent between 1938 and 1952.[2] One analyst estimated that crew wages and benefits plus cargo handling amounted to about 70 percent of total vessel operating expenses. With such an increase in labor costs which made up the majority of the ship operating expenses, financial difficulties could be expected. During the period 1950–52, the ratios of the operating expenses divided by the operating revenues in the intercoastal trade were 104.8, 97.4, and 102.4, respectively. A group of ten representative coastal and intercoastal operators lost $2.3 million in 1953 and $778,000 in 1954.[3] The plight of the domestic operators resulted in a 66 percent decrease in the number of vessels they employed between 1939 and 1955, as shown in Table 6–1. In this climate of desperation, Matson and Sea-Land felt forced to experiment with containerization in an attempt to reduce their high labor costs. Sea-Land began carrying containers in 1956 and Matson followed two years later.

Low Investment Cost

After World War II, the Merchant Ship Sales Act of 1946 made available warbuilt ships to U.S. citizens until January 15, 1951. These ships were sold for cash or on terms. In general, most were purchased on terms, which provided for a 25 percent down payment, 3½ percent interest, and a maturity twenty years from the date of delivery from the shipyard. Table 6–2 shows the prices at which many vessels were sold. In contrast, the cost of the first class of postwar freighters, the Mariners, ranged between $8.25 million and $10 million each. (The subsidized steamship companies acquired these vessels at approximately half the original cost.)

Bosth subsidized and unsubsidized were able to take advantage of the Ship Sales Act. However, by the end of the 1940s most of the subsidized lines had already signed Operating Differential Subsidy contracts with the government. Consequently, although they could convert their present tonnage to containerships, they were now obligated by the ODS contracts to replace these vessels with new construction at the end of their economic life. The unsubsidized lines do not have such obligations, and all their containerships up to quite recently have been converted from warbuilt ships at a relatively low investment.

The unsubsidized lines also obtained many ships from the Vessel Exchange

Table 6–1
U.S. Flag Privately Owned Dry Cargo Vessels, as of June 30, 1939 and 1955

	1939		*1955*	
	No.	*Gross Tons (Thousands)*	*No.*	*Gross Tons (Thousands)*
Total U.S. Fleet	740	3,918.7	822	6,240.0
Domestic Trades*	468	2,170.7	161	1,115.0
Percent Domestic Trade of Total	63.2	55.4	19.6	17.9

*Includes coastwise, intercoastal, and noncontiguous trades.
Source: Wytze Gorter, *United States Shipping Policy* (New York: Harper and Bros., for the Council on Foreign Relations, 1956), p. 138.

Table 6–2
Merchant Ship Sales Act of 1946

Class	*Typical Selling Price*
C-1	$ 970,000
C-2	1,050,000
C-3	1,230,000
Victory - AP3	1,065,000

Source: Robert S. Hope, "Problems of Financing Vessel Replacement," *Merchant Marine Policy,* Marvin L. Fair and Howard C. Reese, eds. (Cambridge, Md.: ©1963, Cornell Maritime Press, Inc.), p. 101.

Program enacted in 1960. With this program, an unsubsidized line could use its present vessels as a trade-in value on vessels in the reserve fleet. The subsidized lines were unable to upgrade their fleets in this manner, but helped the unsubsidized lines by relinquishing their warbuilt ships to the government reserve fleet when new construction was delivered. Although there was no legal bar to subsidized operators selling ships directly to nonsubsidized lines, the practice has been to sell to the government, which resells to the unsubsidized companies, usually at a small loss. As of September 3, 1969, 120 vessels had been traded out while 124 ships had been traded in. Sea-Land, alone, accounted for twenty of the vessels traded out. The Vessel Exchange Program was an important factor in the fleet improvement planning processes of Sea-Land and Matson.

Methodology for Analyzing Fleet Improvement Planning Processes

In the cases of the subsidized lines, their most recent ship designs were studied as the companies entered a changing environment. Sea-Land and Matson have operated in the environment of containerization for more than ten years. Consequently, looking only at their newest designs of full containerships signifies little. Therefore, the general planning processes of both companies were studied from the time they entered containerization. Their ship designs over the years were investigated to discover whether there were changes in these designs to denote weaknesses in the planning processes. The same format used to analyze the subsidized lines was applied more broadly here.

Sea-Land Service

Initial Strategy[a]

McLean Industries, the parent corporation of Sea-Land, was founded in 1955 when Malcom McLean purchased Waterman Steamship Corporation and its subsidiary, Pan-Atlantic Steamship Corporation. McLean started his container service with Pan-Atlantic whose name was changed to Sea-Land Service in 1962. Today Sea-Land, an unsubsidized firm, controls a larger containership fleet than any other privately owned company in the world. The company serves the entire domestic trade as well as the Far East and North Atlantic.

The current container revolution in the ocean trades began with the efforts of Malcom McLean. From his twenty years of experience as the head of McLean Trucking Company, he realized that the overall distribution costs could be reduced only if the whole process of distribution, from shipper's door to consignee's door, was thoroughly streamlined. The problem was not the long-haul operation but the repetitive handling of freight, especially at points where freight was transferred from one mode of carriage to another, for instance at dockside. Here, freight which usually was trucked to the port was loaded by longshoremen onto the ship. Often this longshoring expense at both ends of a voyage in the domestic trades represented more than half the cost of transporting the freight. Realizing this factor, Malcom McLean decided to experiment with containerization in order to reduce the total cost of the origin-to-destination cargo movement.

Planning for the Resultant Ship Designs

McLean started his experiments in the field of containerization in the spring of 1956 with thirty-three-foot containers as deck cargo on tankers running from Houston to New York. Oil was carried below decks in the tankers

[a]For more information on Sea-Land read the case studies, "Sea-Land Service, Inc. (A) and (B)," written by the author in 1970 at the Harvard Graduate School of Business Administration.

while containers were carried on top of the decks. When he realized the potential profitability of moving containers, he changed to full containerships.

McLean did not attempt to set up a permanent planning staff; instead he made great use of his own experience and insight into the total transportation system as well as the knowledge learned by his mistakes to improve his container system. However, he did supplement his steamship company staffing by adding truck line specialists—research engineers, maintenance directors, traffic experts, and others to help plan and direct the container service operations. In addition, in a period of more than ten years, his company spent in excess of one million dollars in review, research, and studies concerning the most desirable container size.

From his initial experiments with thirty-three-foot containers aboard tankers, McLean went to thirty-five-foot containers on full containerships. He had originally thought that roll-on/roll-off would be the proper method of putting the containers aboard the ships. However his experience showed him that the lift-on/lift-off method permitted better use of the vessels' carrying capacity since it did not entail in-transit storage of the chassis along with the trailer.

In 1958, he began a full containership operation in the coastal trades with six converted C-2s having a carrying capacity of 226 thirty-five-foot containers each. In 1962, Sea-Land revived the intercoastal trade with the inauguration of full containership service with four jumbo-ized T-2 tankers converted to carry 476 thirty-five-foot containers each.

Expansion into the U.S. Foreign Trade. In 1964, Sea-Land had expanded as far as it wished in the domestic trades and now contemplated entering the U.S. foreign trade. However, Sea-Land needed more capital before if could purchase or convert more ships.

In October 1964, McLean Industries announced an agreement with a newly formed subsidiary of Litton Industries known as Litton Leasing Corporation. Under the agreement, Litton Leasing purchased six C-4 vessels allotted under the Ship Exchange Act to McLean's subsidiary, Waterman Steamship Corporation. Ingalls Shipyard, a subsidiary of Litton Industries, was designated to refit them as containerships. As they were put into service, the vessels were leased to Sea-Land by Litton Leasing. While the first four ships represented a $46 million reconversion, the agreement also provided for construction of six all-new 905 foot, 1,261 container capacity ships.[b]

Since 1957, Sea-Land had been offering its customers a variety of special-purpose trailers. These included open top, insulated/ventilated, and controlled-temperature trailers. The latter provided exact and constant control from $-15°F$ to $+60°F$.

In 1965, the company introduced two new trailers—a bulk liquid

[b]This provision was later changed, and these ships were not built.

trailer equipped with a 5150-gallon stainless steel tank, and a flatbed trailer, for transporting items such as plywood that require side loading. One hundred and fifty bulk liquid containers were bought at a price of $14,000 each, and were later used in the importing of scotch whiskey on the North Atlantic. By year end, 1965, the company was operating a fleet of more than 12,500 trailers.

On April 23, 1966, Sea-Land inaugurated weekly trailership services to Europe. Four C-2 trailerships, each carrying 226 trailers, were put into this service, calling at Elizabeth and Baltimore in the United States, and at Rotterdam, Bremen, and Grangemouth (Glasgow) in Europe. Seven other European ports were served over-the-road and via commercial ferry connections. Sea-Land thus provided the first all-container service on the North Atlantic.

Sea-Land used a computer control system to keep track of its containers. As the company fleet grew, Sea-Land experienced the benefits of economies of scale in its computer control system, in its marine terminal operations, and in the total number of containers needed for the Sea-Land system. In addition, because Sea-Land had a profitable domestic service in operation before it started in the U.S. foreign trade, it could use the domestic service at a low incremental cost as a feeder system to pick up and deliver containers for its international trade.

In September 1966, Sea-Land had more than one half the cubic capacity of containers owned by U.S. marine carriers, as shown in Table 6–3. Although controlling the largest container fleet in the world, Sea-Land officials, many formerly in the U.S. trucking industry, still maintained an attitude that differed from that of traditional maritime carriers. "We're truckers, not steamship operators," one Sea-Land executive was quoted as saying.[4] The difference may show up in Sea-Land's concentration on offering a door-to-door service—the same tactics used by truckers to beat the railroads on the surface—and on extreme attention to the selection of freight on the basis of revenue, density, and handling costs.

One example of marketing prowess accomplished by Sea-Land and REA Express was the Unit-Pak International service which speeds the routing of liquor imports from Scotland to distributors in the continental United States while reducing transportation costs. The service coordinated the advantages of containerization with the benefits available with palletization. Sea-Land containers loaded with LCL consignments of Scotch whiskey and gin overseas traveled to Sea-Land's Port Elizabeth terminal where the individual shipments were broken down, palletized, and shipped via REA to the customers.

In a separate action, Sea-Land negotiated with the Pennsylvania Railroad to move containers in unit trainloads from East St. Louis and Chicago to the port of Newark located near Sea-Land's headquarters. The tariffs proposed by the Pennsylvania Railroad (which would be available to all shippers)

Table 6-3
Summary of Container Inventory by Number of Containers and Cubic Capacity (Marine Companies Only)* September 1966

Container Length	*Inventory*	*Percentage of Total Inventory*	*Cubic Capacity†*	*Percentage of Total Cubic Capacity*
35 feet	18,865	42.40	39,390,120	52.28
24 feet	6,220	13.98	8,875,152	12.23
Subtotal	25,085	56.38	48,265,272	66.51
10 feet	1,356	3.06	682,146	.94
20 feet	13,450	30.22	14,528,254	20.02
30 feet	0	0	0	0
40 feet	3,001	6.74	6,690,834	9.22
Subtotal	17,807	40.02	21,901,234	30.18
All Other	1,606	3.60	2,402,196	3.31
Grand Total	44,498	100.00	72,568,702	100.00

*Sea-Land owned all the thirty-five foot containers; all the company's containers were of this length.

†Expressed in cubic feet.

Source: U.S. Senate, Committee on Commerce, "Standardization of Containers." Prepared by Subcommittee on Merchant Marine and Fisheries, 90th Congress, 1st sess., July 1967, p. 236.

were based on a flat charge for a trainload which could include up to seventy-five cars. The tariffs were suspended due to protests by ports and railroads who claimed, among other things, that the new rate would greatly undercut the present rate structure.

Expansion into the Far East. In July 1966, Sea-Land began a U.S. West Coast-Okinawa service for the Defense Department. Large-scale containerized shipping to South Vietnam started in July 1967, with a two-year, $70 million contract awarded to Sea-Land. On March 7, 1968, Sea-Land announced that it was engaged in discussions with representatives of Japanese government agencies. Talks with the Ministry of Transport and their Port Director of Yokohama involved suitability of available pier locations, land for marshaling yards for Sea-Land containers, and arrangements for the necessary dockside crane facilities. In December 1968, Sea-Land commenced a weekly eastbound containership service between Yokohama and Seattle and Oakland. These ships carried Department of Defense cargo on their westbound voyage. By early 1969, Sea-Land had expanded its international fleet to six ships in the Atlantic and six in the Pacific.

New Construction. Realizing that most of its competitors had larger and faster containerships either in service or on order, the management at Sea-Land considered the possibility of the construction of as many as eight new containerships. All Sea-Land's present containerships were conversions. All of them were over twenty years old and could maintain speeds of no more than 17 knots. Subsidized competition had under construction more than a dozen ships capable of speeds between 20 and 25 knots, container capacities of 800-1200 twenty-foot equivalent containers, and propulsion plants in the area of 30,000 horsepower.

Sea-Land arrived at a design for the new ships, known as the SL-7s, that offered a capacity of 1,082 trailers of 35- and 40-foot length, 120,000 horsepower, and a service speed of 33 knots. The estimated price in a foreign shipyard would be approximately $32 million per ship, the unusual power plants costing half this amount.

In the spring of 1969, McLean Industries merged with R. J. Reynolds Tobacco Company. This merger provided a needed source of capital from which Sea-Land could obtain funds for new ship construction. Consequently, Sea-Land ordered from foreign shipyards five SL-7s in August and three more in September, for delivery in late 1972 and 1973.

The Changes in the Environment and in the Ship Designs

When McLean started his container experiments in 1956, he was competing against break-bulk cargo carriers as well as various forms of land transportation. Ten years later when Sea-Land expanded into the U.S. foreign trade, the company was well aware of the intricacies of container operations. Since 1956, the size and scope of Sea-Land's operations have greatly increased, and the competition has begun container operations. However, the basic environmental factors affecting Sea-Land's operations have changed little.

Through a sophisticated trial-and-error learning process, Sea-Land changed from using thirty-three-foot containers to thirty-five-foot containers, from carrying containers as deck cargo on tankers to creating full containerships, and from planning roll-on/roll-off ships to building lift-off/lift-on vessels. These changes took place during the initial experimental stages and did not cost the company a great deal of money.

A later change in 1965 involved the installation of dockside cranes. From experience, Sea-Land had discovered that it was better to place its cranes on shore rather than aboard ships for these two reasons: (1) shipboard cranes were unproductive while a vessel was at sea, whereas the dockside crane, which moved on tracks along the berthing channel, could service any number of vessels in succession; (2) because of its greater reach, the elimination of shipboard cranes allowed for an added level of containers above deck.

This change in policy and ship design was less costly to Sea-Land than to a subsidized line. In the first place, all Sea-Land's containerships were conversions from warbuilt vessels. Consequently, they had shipboard cranes on them when Sea-Land acquired them. Second, since Sea-Land owned its marine terminal facilities, it could postpone an investment in dockside cranes while it used the shipboard cranes for loading and unloading.

Sea-Land's informal methods of planning brought prosperity to the company in the late sixties. Sea-Land's success has also been greatly aided by the economies of scale inherent in the firm's expanding operations and the need for vessels in Vietnam. It has been said that in the mid-sixties, it was poor planning by Sea-Land that gave it the necessary ships to accept the $70 million contract for military cargo in the Far East.

Outcome of This Course of Events

By using full containerships and taking advantage of economies of scale, Sea-Land has become the leader in the current container revolution. This achievement has been aided by the financing arrangements of Sea-Land and the need for vessels in Vietnam. The company's full containership fleet of more than thirty vessels is larger than any other privately controlled full containership fleet in the world. In addition Sea-Land presently has under construction or in operation the eight fastest containerships in existence.

By the late sixties, Sea-Land had become dissatisfied with its informal planning process and was establishing a separate group for formal long-range planning. The company planned to use such sophisticated techniques as computer simulation in its future fleet replacement programs.

Weakness in the Planning Process[c]

If there were any major mistakes in Sea-Land's planning process, they were overshadowed by the need for ships in Vietnam. It is important to note that Sea-Land now feels that a formal planning program is necessary since this is quite a change from the informal planning performed by the company in the past.

Matson Navigation Company

Initial Strategy

For many years Matson, an unsubsidized liner firm, has served Hawaii from the Pacific Coast of the United States. In 1967, this company expanded

[c]The author does not wish to leave the reader with the feeling that the planning program of Sea-Land was infallible. Since Sea-Land ordered the SL7s in 1969, their cost to Sea-Land has increased by over 50 percent to more than $50 million per vessel.

its service to include the Far East. Initially, the general stimulus for the Matson research program was the unhealthy economic climate in the domestic trades in the fifties. Saddled with obsolete equipment and an inordinately high labor content throughout its operations, the company was under pressure from the rapidly spiraling wage scale; moreover, Matson felt that it had little or no chance of improving profit margins sufficiently to attract the venture capital needed to correct these conditions. Basically, Matson felt the problem was one of low productivity, but management decided that any financing required to solve this problem would have to come from cost reductions within the company.

In this situation, the first step was to isolate the largest cost reduction opportunity in their operations. This was not difficult since the cargo-handling operations associated with loading and discharging the ships accounted for almost half of the total transportation costs in the Hawaiian trade; in addition, the same costs were responsible for using up almost half of the ship time that would otherwise be available for transportation as such. Consequently, since cargo handling was the area in which cost reduction measures would exert their greatest leverage, Matson focused its attention on containerization.

Matson decided to begin planning in a systematic manner to find the best way of increasing the productivity in its operations. The company's planning process started by trying to discover an optimal system for the particular requirements of the Hawaiian trade. Computer simulation techniques provided the key management tools in analyzing this system. Using these techniques, a large number of hypothetical but likely total fleet situations representing varying degrees of containerization were examined. Selecting a graduated series of the best of these gave Matson its guidelines for a "best" container system and established the implementation phases through which they wished to progress. Each increment of capital required in this planned series of events was evaluated for its potential to generate sufficient return (cost reduction) to justify the investment.

Resultant Vessel Designs

In 1954, the president of Matson gave the order to organize a suitable research activity aimed at improving cargo transportation and distribution systems. Initial surveys had indicated that there could be substantial payoffs to scientific research analyses of transportation methods. Although sea-going equipment had been considerably improved as new materials or power sources were discovered, the basic system of handling, gathering, and distributing cargo was essentially unchanged since the time of the Phonecians. In 1955, the first research people were hired. The next year, an internationally known consultant firm was hired to assist in setting up a full-scale operations research department. The consultants advised that a staff of operations research people should be employed and that Matson could profitably spend $750,000 per year for research.

Stanley Powell, the president of Matson stated in congressional hearings,

> We did not hire any steamship people in this research activity but, rather, hired experts in systems analysis. We felt that if we got people who were not knowledgeable in the steamship business, they would not be impeded in their thinking by the acceptance of things that had happened to be there for many, many years.
>
> In 1957, an engineering development department was formed to work in concert with research on engineering and equipment design problems. Again we hired professional people. We did not put ex-chief engineers or skippers in here, but people who had professional and educational backgrounds and experience. As for as I know, Matson is the only steamship company in the world to have this kind of research and engineering team constantly at work studying improvements in transportation systems.[5]

Computer Simulation Model.[d] The development of a computer simulation model as an aid in the solution of operations problems was instituted by Matson in the late 1950s after a careful investigation of its feasibility. The "traditional" method permitted a limited number of possible fleet replacement programs under fixed circumstances to be evaluated. The model conceived by Matson represented a logical step forward from the "traditional" method since it could evaluate an unlimited range of possible replacement programs under varying circumstances. Further, the model could provide quantitative information to help with many other vital operational problems.

Matson developed the model with the following objectives:

1. To develop detailed operational cost comparisons for the major commodity-ship-port combinations of interest in Matson's freight operations.
2. To develop tactical rules for selecting the minimum-cost mode of operations under any likely set of conditions regarding cargo and ship-type availability.
3. To estimate the probable change in net freight revenue that would result from any combination of the following changes in operating conditions:
   a. Changes in total cargo mix, such as would result from expansion of such container service, or from expansion or reduction of other major cargoes such as lumber, sugar, or pineapple.

[d]Information on Matson's computer simulation model has been taken from "Operational Simulation of a Freighter Fleet," by Foster Weldon, in *Research Techniques in Maritime Transportation,* by NAS-NRC, Maritime Cargo Transportation Conference, Washington, D.C., 1959, pp. 21–8, and "Trade-Fleet Study" by NAS-NRC, Maritime Cargo Transportation Conference, Washington, D.C., 1960, pp. 14–21.

b. Changes in ship scheduling and ports of call.
c. Changes in ship types and fleet composition, such as the introduction of container ships, specialized bulkers, etc.
d. Changes in labor practices.

4. To assess the implications of objectives 1, 2, and 3, with respect to development of a logical tariff structure and an *optimum fleet replacement program.*

Outputs. There were four types of output from the model. These were the profit and loss statement, the comments tape which presented miscellaneous operational information, the manifest, and the voyage report.

The profit and loss statement was a Consolidated Profit and Loss Statement for the operating period covered by the model (one year) and reported the profit or loss sustained as a result of each simulation. This output was of greatest interest to management because it told the possible consequences, in dollars and cents, arising from each simulation.

The comments tape presented varied operational information. It showed the incidence of port congestion, identified commodities which had built up to critical (excessive) levels, and told the total number of containers needed for the fleet operation. This information was of great use to the company. For example, the incidence of port congestion reflected the adequacy of the schedules being followed, while the record of the number of containers required showed the adequacy of the container supply on hand during the course of the simulation.

The manifest was a detailed listing of cargo carried per voyage by each ship, broken down by commodity; ports of origin, destination, and transshipment; and packaging mode. It also detailed such individual voyage data as cargo handling manhours, days at sea and in port, and days delayed or in lay-up. This detailed information was for determining the individual performance of each ship in the fleet.

The voyage report was a summary of the information presented on the manifests and enabled an evaluation of the overall fleet performance. Further, it related fleet performances to the profit and loss statement.

Model Structure. The Matson model was a simulation of Matson's freighter fleet operations between two "sites," the west coast of the United States and the Hawaiian Islands. It enabled the operations to be simulated for the period of one year, using the inputs and providing the outputs described earlier. The model handled the inputs in three general stages: (1) a cargo generator; (2) an operations model; and (3) an accounting model. The cargo generator, utilizing the cargo data input, told how much cargo of each type was present, on a daily basis, at each of the ports in the two sites. The operations model, utilizing the operations input and the outputs of the cargo generator, made all the detailed calculations that were involved in controlling the operations of the fleet at sea and in port. The accounting model, using the accounting data input and the results of the operations model, turned out the final balance sheet telling how well the company had performed financially during the year.

Model Flexibility. Flexibility in the model, a product of careful design, was achieved by varying the following inputs to reflect changes in operational situations.

1. Fleet Mix and Vessel Characteristics. The model permitted consideration of ten classes of vessel (Mariner, C-2, C-3, etc.) and a total fleet of twenty-five ships made up of these ten classes. By permitting consideration of these ten classes, it was possible to obtain information concerning the effects of special-purpose ships (container, break-bulk, etc.) upon the Matson trade.

Further, the characteristics of the twenty-five ships could vary. No one ship needed to be identical to another in all respects; e.g., cargo space configuration could vary. For example, one Mariner-class vessel could have 700,000 bale cubic feet of dry cargo space and another, 500,000 bale cubic feet of dry cargo space, with a consequent increase in some other type of cargo space. Thus, the model enabled many fleet mixes to be considered and the best selected as the basis for the fleet replacement program.

2. Competition. One of the basic points of interest was the manner in which the model considered competition. Of the fleet of twenty-five vessels that the model permitted, any or all of them could be designated "Matson" or "Enemy." All were treated in a similar manner and, generally, under the same rules. Also, cargo offerings could be biased to favor Matson or its competition. Thus, by varying the number of "Matson" and "Enemy" ships in the fleet and by varying cargo bias, a quantitative measure of the effects of stable and changing competition on Matson trade could be ascertained.

3. Fleet Operations. The model did not require that all of the fleet be in operation simultaneously. Decision rules had been established and incorporated into the model which determined when ships would be laid up or additions would be made to the fleet. The major criterion for additions or lay-ups was the amount of cargo available to ships on the schedules that they might follow. This criterion could be changed to reflect various situations.

4. Cargo Handling Methods. The model considered the handling of general cargo by two methods, containerized and break-bulk.

Man-hour productivity rates had been established for containerized and noncontainerized commodities. These could be altered to reflect improvement in cargo handling techniques.

There was an established order in which cargo would be loaded, i.e., reefer, containerizable cargo, dry cargo, deep tank, etc. This order had been established to minimize loading time and to expedite shipment of high priority cargo. It could also be varied to determine the effect of loading sequence upon operations.

5. Schedules. The model permitted consideration of all schedules fed into it. Schedules were prepared outside the model for input purposes on the basis of an examination of the possible port-pairs, cargoes available, and the available ships. These schedules were then reviewed and those selected were placed in the model for consideration by the fleet. This selection could be changed as desired.

It should be noted that all ships could not consider all schedules; each ship was identified with schedules that it could select. For example, high-speed vessels might have been identified with schedules containing ports shipping high priority commodities. By changing the ship-schedule identification, it was possible to determine the consequences of changes in ship routing upon operations. Further, by changing the entire selection of schedules placed in the model for consideration by the fleet, it was possible to determine the effectiveness of various routing patterns.

Once a ship had selected a schedule from among its alternatives, the model still allowed changes in the vessel's course of action. Schedules were subject to evaluation periodically to determine if they were still optimum. It was recognized that the conditions prevailing at the time the schedule was selected might not exist for its duration. Thus, an evaluation was made just prior to the conclusion of each leg of the schedule. At this time, the remainder of the prevailing schedule was compared to other alternative schedules available to the ship. If it was found that an alternative was superior to the present schedule, it would replace that schedule.

In order to prevent a ship from replacing a schedule with one that would cause it to retrace its steps, schedules were route-oriented, as well as ship-oriented. That is, once a ship selected a schedule on a given route, such as San Francisco-Portland-Honolulu-San Francisco, it could not select another schedule that would take it off this route through the Pacific Northwest. Using this route as an example, the vessel could not be diverted at Portland to Long Beach; only after it had called at Honolulu and returned to San Francisco from the Hawaiian Islands could it be diverted to Long Beach. The model did not permit *intrasite* shipments. Only *intersite* shipments were allowed.

This leads to the question of the disposition of cargo on shipboard destined for a port which was no longer scheduled. In this instance, the cargo was unloaded at a scheduled port within the same site as its port of destination, for transshipment to its destination. In cases where a ship was dropped from the active fleet due to lack of trade while still having cargo aboard, this cargo was discharged at an "unloading" port within the site (a port designated for this purpose); the ship was then withdrawn from the fleet until recalled.

6. Cargoes. The model permitted a large number of commodities to

be considered and also seasonal variations and fluctuations in commodity mix. Further, it was possible to eliminate a commodity from consideration during the simulation if it was felt that this would reflect a "true-life" occurrence. For example, it might have been that certain commodities were shipped during only a portion of the year. The model would enable consideration of this commodity for that period, and then eliminate the commodity from further consideration.

The Planning Process[e]

With the aid of its computer simulation model Matson planned for the transition to a fleet of high-productivity vessels. Three years of research and engineering went into the initial design of the Matson containers before one was built to the Matson standard dimensions of twenty-four feet long, eight feet wide, and eight feet, six inches high. Unlike other companies, Matson never used a containership with shipboard cranes. Their research had showed them the advantages of having the cranes on shore.

Phase One

Matson's container service started on an inexpensive pilot basis in August 1958. Matson outfitted conventional general-purpose vessels to carry the containers as deck loads only. Initially forty containers were handled, but the number gradually increased to almost one hundred per voyage on six different vessels. Cargo handling cranes were mounted on the docks to handle the containers. This first phase represented the lowest level of container operations that would provide a useful test of service. At the same time, it was the highest level of financial risk Matson wished to take, considering the possibility that the longshoremen might have refused to cooperate with the work-saving innovation.

Phase Two

During this first experimental phase of container operations, plans were finalized for converting the C-3 type vessel *Hawaiian Citizen,* to a full containership with cellular holds and a container capacity of 436 containers. By the fall of 1959, Matson felt that the labor risk associated with the container-

[e]The description of the Matson planning process is based on the paper "Research in Steamship Operations" by Foster L. Weldon, in Marvin L. Fair and E. Grosvenor Plowman, eds. *Coordinated Transportation, Problems and Requirements* (Cambridge, Md.: © 1969, Cornell Maritime Press, Inc.

ization had been reduced to an acceptable level, and the conversion work was started. The *Hawaiian Citizen* went into service in May 1960.

Phase Three

At the time a firm decision was made to convert the *Hawaiian Citizen,* Matson was serving Hawaii with break-bulk cargo vessels which carried general cargo westbound and carried bulk sugar eastbound. Because of the high productivity of the *Citizen,* the freight rate on break-bulk cargo would be reduced when the converted vessel went into service. As long as the bulk sugar traveled in unconverted cargo ships, there would be no lowering of its freight rate. If the westbound cargo on sugar voyages could be put in containers carried on the same ships, the productivity of the vessels carrying sugar would be greatly increased and the sugar would get a substantial benefit. Consequently, phase three resulted in jumbo-izing two C-4s, placing hopper-hold midsections in them. Containers could be carried on deck as well as in the hold on the outbound leg. In this way, up to 390 loaded containers could be carried outbound, or westbound, and 16,000 tons of bulk sugar and a deck cargo of containers could be carried inbound, or eastbound. The balance in container flow was maintained because a large percentage of the inbound containers were empty, and the sugar provided the stability necessary for higher than normal stacking of this lighter container load entirely on the deck on the eastbound voyage. These two ships, the *Hawaiian* and the *Californian,* were placed into operation in the middle of 1960, greatly reducing the eastbound rate on sugar. In this way, Matson appeased the sugar industry which had by this time figured out that they could buy jumbo-ized T-2 tankers and ship their sugar in them at a rate lower than Matson's previous rate.

Phase Four

Also at this time, the automobile exporters were threatening to provide their own transportation. Again it was clear that specialized transport was possible at a cost appreciably lower than that for the conventional method. Since Matson could not lower their automobile rate without changing their method of auto carriage, the *Hawaiian Motorist* was designed to carry up to 519 automobiles below decks plus 186 containers above decks. The basic secret of the *Hawaiian Motorist's* performance was the use of the automobile's own motive power in loading and unloading the ship. A car was simply driven into an open-ended frame at shipside, hoisted into the hold with the ship's tackle, and then driven out of the frame into a parking stall in the ship. With this method, the ship had a port turnaround time of less than a day.

This obvious procedure had never been used on ocean-going freighters before because of a Coast Guard regulation that prohibited mixing hazardous cargo with other cargoes in the same compartment of the ship. Consequently, in conventional operation, cars had to be immobilized by draining their gasoline and removing their batteries before they could be placed in the hold. When Matson's engineers asked for authorization to operate routinely with nothing but hazardous automobiles in the hold, the Coast Guard approved, and the *Hawaiian Motorist* went into service in December 1960.

More Recent Vessels

Since 1960 Matson has added more vessels to its fleet with capabilities for carrying containers and/or automobiles. A container service was also started between the Hawaiian Islands with a small, specially designed vessel. In 1967, Matson expanded into U.S. foreign trade, becoming the first company to start a commercial full containership service to the Far East. This growth allowed Matson to take advantage of economies of scale in containerization and to use its domestic fleet as a feeder service for its international trade. Matson has decided to build four full containerships of a new design capable of carrying 1168 twenty-four foot containers at a service speed of twenty-three knots. Two ships will be used in the domestic trades while the other two will be employed in the U.S. foreign trade.

Changes in the Environment and the Ship Designs

Since Matson applied its system-analysis approach to an environment which could include containerships as well as break-bulk and ro-ro vessels, the company has been able to anticipate all the changes in its environment. Matson has been able to apply sophisticated operations research techniques successfully to its fleet improvement planning process and has not needed to change any of its ship designs after the initial experimental stage.

Outcome of This Course of Events

Matson has grown and prospered without the need to change any of its ship designs. From 1961 to 1967, Matson improved from a loss of $1,067,100 to a profit of $9,460,000.

Weakness in the Planning Process

With careful planning Matson has avoided making any discernible mistakes in its planning process. It has successfully anticipated changes and future needs in its environment and has correctly predicted the beneficial effects of each phase of its planning program.[f]

[f]The author does not wish to leave the reader with the feeling that the planning program of Matson was infallible. In 1969, finding that its service on the route between the United States and Japan was uneconomical, Matson filed to receive operating differential subsidy. Matson did not receive ODS and terminated this international service in July 1970.

7 Analysis of Planning Processes: Subsidized and Unsubsidized

A major reason for the blundering of some of the subsidized lines through the transition period to the widespread adoption of containerization must lie with their planning processes. Because the subsidized lines had a vested interest in maintaining the past methods of cargo handling and since the lines had little need for long-range planning during the break-bulk era, these companies did not emphasize formal planning and research. The personnel instrumental in planning ship replacement once every several years were normally employed in the day-to-day operation of the company. Once the stable environment had changed, these procedures were no longer adequate, as demonstrated by the studies in Chapter 5. These planning processes and those of Matson and Sea-Land are summarized in Tables 7-1 and 7-2.

The Subsidized Lines

States steamship company performed only a superficial analysis of its environment. The management realized that containerization was coming. However, they had no desire to spend additional money on market research. The company decided that it did not want to take the financial risks of going into large-scale containerization and felt that it could continue to practice the traditional methods of operation in its changing environment. The company apparently later decided that its wishful thinking was mistaken. Consequently, States decided to build roll-on/roll-off vessels at about the time it was receiving brand new break-bulk cargo ships. If States had placed more emphasis on performing a thorough initial analysis and updating it, the company might have a brand new flight of more valuable ships today.

American Mail Lines finally decided to convert some C-4s to full containerships at about the time it was receiving delivery of brand new break-bulk cargo ships. The company could have converted some of their new vessels to full containerships during the construction period, if more emphasis had been placed on expanding and updating the initial analysis.

Professor Paul W. Cherington summed up the plight of companies such as States and American Mail Lines in avoiding large-scale containerization:

> Sure, there are financial risks. But there's really no choice for shipping lines. It's analogous to the father who is asked to spend $5,000 on hospitalization for a sick child. If he spends

Table 7–1
Summary of Subsidized Lines Planning Processes

	States	*American Mail Line*	*Moore-McCormack*	*Executive Lines*
Initial* Strategy	Follow competitor's design and hope for success in a changing environment by continuing to operate following the methods of the status quo.	Continue as break-bulk cargo operator after deciding that large-scale containerization would not come to its trade route.	Be ready to meet any changes in the environment.	Be flexible for the transition to containerization.
Initial Design	Break-bulk cargo ship carrying 216 containers.	Huge break-bulk cargo ship carrying 411 containers.	Very flexible ship carrying 832 containers or break-bulk or ro-ro cargo.	Unique break-bulk cargo ship carrying 228 containers.
Outcome	About the time new break-bulk cargo ships were arriving, States decided to build full containerships, later changed to ro-ro vessels.	About the time new break-bulk cargo ships were arriving, AML was trying to convert other break-bulk ships to full containerships.	Has decided to sell ships after they were delivered and withdraw from North Atlantic trade.	After two changes in design, constructed a 1200 container capacity full containship-unnecessary expense of more than $3 million and time delay of more than one year.
Weakness in the Planning Process	Wishful thinking and superficial analysis are not sufficient in a changing environment.	Failed to evaluate thoroughly the possible impact of inland transportation and the economic consequences of full containership competition.	Underestimated the costs of flexibility in a specialized market.	Concentrated on the immediate situation rather than long-range picture.

*Refers to the original rationale in the most recent flight of ships for each company. These also were the first attempts at containerization on the part of these four companies.

Table 7–2
Summary of the Planning Processes of Sea-Land and Matson*

	Sea-Land	*Matson*
Initial Strategy	Use containers to reduce cost of origin to destination cargo movement.	Minimize costs for total cargo movement in the system.
Resultant Designs (after experimentation)	Full containerships with ship-board cranes.	Full containerships using dockside cranes as well as unique combination designs.
Outcome	Financial success, only design change was from shipboard cranes to dockside cranes in 1965.	Financial success, no design changes.
Weakness in Planning Process	If there were any major mistakes, they were overshadowed by the need for ships in Vietnam.	No major mistakes in the planning process.

*"Initial Strategy" in this exhibit refers to the original attempts of Sea-Land and Matson at large-scale containerization, rather than their strategy for their most recent ship design.

> the \$5,000, his child may wind up no better off than he started out—but if he doesn't agree to spend it, he may not have a child. If you're not ready to get into containerization today, you may not be here tomorrow.[1]

Both Moore-McCormack and Executive Lines planned their last flight of ships while concentrating on the transition period between break-bulk cargo ships and full containerships. Moore-McCormack managed to construct some very flexible ships. However, the company underestimated the cost disadvantage of flexibility in their new ships as well as in their general operations. Moore-McCormack realized the high cost of maintaining flexibility when it was forced to compete against specialized containerships on the North Atlantic. Eventually, the company saw no alternative to selling its new flexible vessels and withdrawing from this trade route. If Moore-McCormack had put more emphasis on its long-range strategy and the economic consequences of a high degree of flexibility, changes would have taken place, both in the design of the vessels and in the company's general operations.

Executive Lines managed to build some very productive ships but only after changing its design twice and concentrating on economic considerations in the long run. This procedure cost them an added expense of more than \$3 million and a delay of almost two years. If Executive Lines had placed more emphasis on long-range planning initially, it would have been able to perform its ship conversions with less cost and delay in one of two ways. Either the initial design would have possessed the ability to be converted to a full containership without requiring a complete redesign, or the initial conversion would have produced the final design.

Unsubsidized Lines

Both Sea-Land and Matson concentrated on the cost of the door-to-door rather than the port-to-port movement of cargo. Sea-Land used a crude systems analysis approach. However, the company was able to recognize its errors and respond quickly to solve them. If Sea-Land made any mistakes in planning by growing too fast in the mid-sixties, the need for ships in Vietnam disguised them.

When Sea-Land first entered the U.S. foreign trade, it was the only company offering full containership service. Since that time, many other liner firms have adopted this practice. With its competitive situation changing, Sea-Land felt that it was important to have a permanent planning staff. The company decided to establish one and planned to use such sophisticated techniques as computer simulation.

Matson was the first U.S. steamship line to use formal systems analysis techniques in its planning process. By setting up a separate research staff,

Matson has been able to apply computer simulation techniques to the company's entire operation, including an allowance for actions of competitors. By means of painstaking analysis, this unsubsidized firm has been able to predict the needs of its particular trade route; moreover, as a result of its planning exercises, Matson has designed and built such unusual vessels as a combination container-bulk sugar carrier and a combination container-car carrier to take advantage of the particular needs of its operating environment. By the end of 1967, Matson was spending more than $1 million per year on research. In that year, the company showed a profit of $9,460,000, up from a loss of $1,067,000 in 1961. Careful planning appears to have paid off for Matson.

Use of the Computer

Computer simulation techniques used by Matson have the advantages of: (1) capacity to handle a great number of variables, (2) speed, (3) accuracy, and (4) adaptability to handle a great number of problems simultaneously. Their major disadvantages are: (1) cost and (2) the lead time required to accumulate necessary input and to design the simulation. In 1960, the National Academy of Sciences felt that the amount of funds involved in vessel replacement programs warranted investigation of the advantages of simulation techniques by the subsidized lines.[2] In the past ten years, the cost of ship replacement has greatly increased; consequently, the author feels that the advantages of computer simulation are even greater today than at the time of the National Academy of Sciences' study. It must be emphasized that computer simulation is a tool, an aid to management in decisionmaking. Simply using a computer cannot solve the planning problems of the company. However, based on the inputs supplied to the simulation model, the computer can provide management with quantitative information concerning the probable consequences of alternative courses of action. The final decisions must rest with management. Other factors which affect the success of the planning process are the attitude of management and the flexibility of the company in its operations.

Attitude

The planning processes which Sea-Land and Matson used in their systems analyses do not fully explain the successful innovations of these two unsubsidized firms. "Systems analysis is as much an attitude—a philosophy—about exploring a many-dimensional problem as it is a collection of sophisticated techniques."[3] Without this orientation, the technology of containerization and the techniques of operations research would have been applied piecemeal, as the subsidized lines have done. The difference in the two segments of the industry has been basically one of attitude.

This difference became manifested in the personnel whom the two companies entrusted with executing the new systems. Rather than having management made up solely of veterans in maritime administration, both unsubsidized lines adapted management concepts from other areas. When Malcom McLean, the founder of Sea-Land, purchased Waterman Steamship Corporation and its subsidiary, Pan-Atlantic Steamship Corporation, he supplemented their staff with truck line specialists, including research engineers, maintenance directors, traffic experts, and others. When Matson formed a research group, the firm purposely hired experts in operations research rather than experienced maritime personnel, so that the conventional practices of the steamship industry would not inhibit the formation of new ideas.

During Congressional hearings Stanley Powell, the president of Matson, drew the following contrast:

> I was shocked by a statement that one of the previous witnesses (from a subsidized line) last week made when he said:
>
> > After all, there is nothing that we are doing that the Japanese cannot do, or our foreign competition cannot doand keep up with us, and then we will need construction subsidy and operating subsidy in order to keep up with them.
>
> Well, the way to keep up with them, I think is to keep 10 years ahead of them at all times and not assume that once this little containerization bit is finished, that is the last bit of innovation that is available to the U.S. merchant marine.[4]

Others have been more caustic in their evaluation of subsidized management. One industry observer commented:

> The only way you are going to solve the problem is by bringing in new capital and new management. The merchant marine industry today is the product of five generations of incest; sons and sons-in-law inherit firms. The industry has never known anything but subsidized shipping.[5]

The progressive attitudes of Matson and Sea-Land were caused and aided by factors unique to the unsubsidized lines. The threat of financial ruin in the domestic trades forced Sea-Land and Matson to look at new methods of moving cargo. The protection of the domestic trades and the ability to obtain containerships with low investments encouraged innovation. The freedom from government regulations allowed Sea-Land and Matson great flexibility and permitted expansion of their operations to take advantage of economies of scale.[a]

Hopefully, any needed change in the attitude of subsidized management will be brought about by the changing factors in its environment. The threat of economic disaster in the future and pressure from profit-oriented conglomerates in the maritime industry both act to encourage progressive attitudes. However, government regulations still act as a hindrance to innovation for the subsidized lines.

Disadvantages of Subsidy[b]

During the sixties, the subsidized lines had three main advantages over the unsubsidized companies: construction differential subsidy, operating differential subsidy, and voluntarily deposited tax-exempt funds for ship construction. With CDS funds, the subsidized lines can build in U.S. shipyards while paying only the foreign shipbuilding cost. For the unsubsidized lines to construct new ships in U.S. yards would cost them about twice as much as if they went to foreign shipyards. The unsubsidized lines do have the option of building overseas; however, these ships cannot be used in the domestic trades. In addition, even if these vessels fly the U.S.-flag, for three years they cannot carry the 50 percent of government-sponsored nondefense cargo reserved for U.S. bottoms.

The ODS funds are given to subsidized lines to put their vessels' operating expenses on an equal basis with those of foreign lines. For a Mariner class vessel, these funds can amount to $700,000 per year. Unsubsidized lines do not receive this form of government aid.

The third major benefit the subsidized companies enjoy is the ability to make voluntary deposits into a special reserve fund. This money is normally tax-exempt if it is spent on ship construction, ship repair, containers, lighters, or ship conversions.

While receiving the forms of government aid described above, the subsidized liner firms did not possess the following advantages enjoyed by the unsubsidized lines in their innovations: (1) unrestricted movement of ships, (2) unrestricted acquisition of vessels, and (3) unrestricted financial operations.

---

[a]Various government agencies regulate the freight rates of all U.S. common carriers, both subsidized and unsubsidized. The regulations referred to here are those affecting the company's ability to operate its equipment and its finances as it wishes.

[b]This chapter first studies the regulatory environment as it existed in the sixties when the many companies analyzed were replacing their ships. Later in the chapter, changes brought about by the Merchant Marine Act of 1970 are described.

Government Attitude

The Maritime Administration, under delegation of authority by the Secretary of Commerce, is responsible for administering the subsidy program. Since the annual Maritime Subsidy expenditures have exceeded $200 million in recent years, as shown in Table 7-3, the Maritime Administration feels that it has an important duty in ensuring that these funds are spent wisely.

Table 7-3
Maritime Subsidy Expenditures ($millions)

Fiscal Year	*Construction Differential Subsidy**	*Operating Differential Subsidy*	*Total*
1956	16.0	135.3	151.3
1957	20.3	108.3	128.6
1958	27.3	120.0	147.4
1959	28.7	127.7	156.4
1960	74.0	152.8	226.7
1961	103.3	150.1	253.5
1962	141.0	181.9	322.9
1963	94.7	220.7	315.4
1964	78.9	203.0	281.9
1965	87.7	213.3	301.0
1966	73.4	186.6	260.0
1967	82.5	175.6	258.2
1968	97.7	200.1	297.8
1969	95.5	194.7	290.2
1970	75.0	205.7	280.7

*Includes funds for reconstruction

†Total may not equal CDS plus ODS because of the rounding off of numbers.

Source: U.S. Dept. of Commerce, "MARAD 1971" (Washington, D.C.): Government Printing Office, p. 25.

How does the Maritime Administration fulfill its responsibilities within the total integrated transportation concept? In the field of intermodal containerization the answer is clear. For example, all new replacement ship designs are reviewed for economy of operation and operational feasibility. Traffic experts study the cargoes to be carried to the routes to be served. They determine the susceptibility of the commerce to unit-handling techniques. They consider traffic projections and examine the status and potential of inland transportation

> services and handling techniques. They study the relationship of traffic administration and shipping practices to the proposed means of operation. Naval architects and engineers study ship design in detail as it relates to traffic and areas to be served. They analyze the appropriateness of the proposed cargo handling gear, container capability and other technical aspects of a planned vessel. The adequacy of ancillary operations and hardware such as terminal equipment, marshalling areas, and rail and highway capacity are considered. In effect, a proposed ship is studied from stem to stern in relationship to the system within which it is to operate.[6]

Since the government considers itself responsible for examining every detail of subsidized operations, it is not surprising that the subsidized industry feels that the government has usurped part of its management functions.

> Somewhere along the line the idea has mistakenly arisen that the Government must act as a guardian of the steamship industry and that companies should never be completely free to exercise ingenuity or imagination in their business decisions. No industry can operate productively or progressively in such a straitjacket. Intelligent management decisions are impaired, if not nullified, when they must be filtered through a mass of governmental bureaucracy.[7]

Trade Routes and Sailing Frequency

Government regulations affect subsidized management's ability to choose trade routes and sailing frequencies. As of June 30, 1968, the subsidized lines provided service on twenty-nine different trade routes as shown in Table 7–4. Trade routes 5-7-8-9 are actually considered as one trade route on the North Atlantic since any company running on one of these routes provides service on all four. Consequently, there are really twenty-six different trade routes being used. Each subsidized line provides service on an average of 2.6 trade routes.[c] On each trade route the government determines the minimum and maximum number of sailings allowed to each subsidized line.

Meanwhile, the unsubsidized lines are not restricted in their trade routes, schedules, or frequency of sailings. This means that the unsubsidized fleets can move their operations to certain areas to meet peak demands whether these conditions are caused by commercial factors or movement of government cargo restricted to U.S. bottoms. Unlike subsidized lines, they can charter vessels as they wish. Consequently, if they want to enter a new trade route

---

[c]Grace Lines and Prudential Lines are considered as two separate companies for this calculation.

Table 7-4
Service on Essential U.S. Foreign Trade Routes as of June 30, 1968

Operator	*Trade Route*	*Annual Sailings Min:/Max:*	
American Export Isbrandtsen Lines, Inc.	U.S. North Atlantic/Mediterranean (T.R. 10), Passenger	24/31	
	U.S. North Atlantic/Mediterranean (T.R. 10), Passenger	10/13	
	U.S. North Atlantic/Mediterranean (T.R. 10), Freight	76/102	
	Great Lakes/Western Europe	6/9	
	Great Lakes/Mediterranean	9/13	
	U.S. North Atlantic/Western Europe (T.R. 5-7-8-9)	6/10	
	U.S. North Atlantic/India-Pakistan (T.R. 18)	24/29	
	Round-the-World Eastbound	24/28	
	U.S. Atlantic/Far East (T.R. 12)	24/30	
	U.S. North Atlantic/Continent (T.R. 5-7-8-9)	18/26	
American Mail Line Ltd.	U.S. Pacific/Far East (T.R. 29-17-28)	36/46	
American President Lines, Ltd.	California/Far East (T.R. 29), Passenger	23/27	
	California/Far East (T.R. 29), Freight	32/48	
	Round-the-World (Westbound)	24/36	
	Atlantic Straits (T.R. 17) (Atlantic–California/Far East-Indonesia and return).	12/28	
Delta Steamship	U.S. Gulf/East Coast South America (T.R. 20)	43 min.	overall max.not to exceed 79
	U.S. Gulf/West Africa (T.R. 14)	24 min.	
Farrell Lines Inc.	U.S. Atlantic/West Africa (T.R. 14)	36/48	
	U.S. Atlantic/South and East Africa (T.R. 15A)	24/30	
	U.S. Atlantic and Gulf/Australasia (T.R. 16)	16/21	

Table 7-4 (cont.)

Operator	*Trade Route*	*Annual Sailings Min:/Max:*
Grace Line Inc.	U.S. Atlantic/West Coast South America (T.R. 2)	96/106 (Comb. & Freight).
	U.S. Atlantic/Caribbean (T.R. 4)	96/119 (Total). (48/53 Pass. Comb). (24/50 C2 Comb.)
	U.S. Pacific/East Coast South America (T.R. 23, 24, 25)	24 min. overall max. not to exceed 60
	U.S. Pacific/Carib. & W. Coast Central America & Mexico (T.R. 23,25).	20 min.
Gulf & South American SS. Co., Inc.	U.S. Gulf/West Coast South America (T.R. 31)	30/36
Lykes Bros. Steamship Co., Inc.	Gulf/Mediterranean (T.R. 13)	42/48
	Gulf/South and East Africa (T.R. 15B)	18/24
	Gulf/Caribbean (T.R. 19)	20/30
	Gulf/U.K.-Continent (T.R. 21)	90/110
	Gulf/Far East (T.R. 22)	48/60
Moore-McCormack Lines, Inc.	U.S. Atlantic/East Coast South America (T.R. 1), Comb-	19/23
	(including: U.S. Atlantic/ Scandinavia (T.R. 6)	(1/4)
	U.S. North Atlantic/E. Coast S. America & S. and E. Africa	(1/2)
	U.S. Atlantic/East Coast South America (T.R. 1), Freight	84/96
	U.S. Atlantic/Scandinavia Belgium & Netherlands (T.R. 6 and 5-7-8-9), Freight.	54/66
	U.S. Atlantic/South and East Africa (T.R. 15-A), Freight	24/30
The Oceanic Steamship Co.	U.S. Pacific/Australasia (T.R. 27), Freight	10/13

Table 7-4 (Cont.)

Operator	*Trade Route*	*Annual Sailings Min:/Max:*
	California/Australasia (T.R. 27), Combination	12/16
Pacific Far East Line, Inc.	California/Far East (T.R. 29), Freight	53/63
Prudential Lines, Inc.	U.S. North Atlantic/ Mediterranean	28/35
States Steamship Co.	Washington-Oregon/Far East	10/16
	Washington-Oregon-California/ Far East	20/30
	California/Far East	22/28
United States Lines, Inc.	New York/Europe (T.R. 5-7-8-9) Passenger	21/25
	U.S. North Atlantic/Europe (T.R. 5-7-8-9)	235/269
	U.S. South Atlantic/Europe (T.R. 11)	34/42
	U.S. Atlantic/Far East (T.R. 12)	45/55
Total 14		1,622/2,038

Source: U.S. Department of Commerce "1968 Annual Report of the Maritime Administration," (Washington, D.C.: Government Printing Office, 1968), pp. 76-77.

with little investment, the extra ships needed can be chartered. As the unsubsidized lines continue to expand their operations to reap the benefits of economies of scale, they are free to serve any or all of the twenty-six different trade routes, while the subsidized lines are severely restricted.

Obtaining New Vessels

Government procedures can be particularly harmful in acquiring new ships. During a period of transition, it is essential that a company have the proper type of vessels to meet the competition. However, when many subsidized lines are crying for new vessels at the same time, the government programs become less able to respond quickly. Before the Maritime Subsidy Board allows a contract with CDS to be awarded, it must conduct both technical and economic studies. Government agencies may propose twelve or thirteen hundred changes before the vessel design is approved.[8] An executive of a subsidized line who was experienced in dealing with Maritime Administration personnel at all levels remarked to the author, "Where else can you find a $9,000 a year employee who can hold up a multimillion dollar investment?" Moreover, not only must a company wait to have its design approved but it must also compete against other subsidized lines for limited CDS funds. Even after the contract is awarded, construction may be delayed until the shipyard has completed some of its previous backlog of work. The subsidized line must wait longer for its ships to be built in a U.S. yard than its foreign competitor does purchasing his vessels in the world market. Consequently, the subsidized line has little incentive to innovate by building a new ship design when a foreign competitor can copy this design and have a vessel in service before the subsidized ship is launched.

The unsubsidized lines can get quick delivery of their vessels by either using foreign shipyards for new construction or U.S. yards for conversions. For example, Sea-Land recently completed a containership conversion to a capacity of 622 thirty-five foot containers. Less than a year elapsed for contract negotiations, vessel design, model testing, detail engineering, shipyard fabrication, erection, and testing time.[9] Subsidized lines making applications to the Maritime Subsidy Board for either conversions or new construction have waited more than twice the time just to get a contract awarded. Prudential Lines, the first company in the world to decide to build LASH design vessels, waited about three years before the government awarded a construction contract. Prudential waited about one and three-quarters years for the keel to be laid. After almost five years of waiting, they had little to show for their efforts. However, while this subsidized line was still waiting for its ships, another line had received delivery of a LASH vessel from a foreign shipyard, although it contracted for construction at a later date than Prudential.

During the sixties, the government did not accept any subsidized vessel less than seventeen years of age as a trade-in value on new subsidized construction. Consequently, if a subsidized operator wanted to innovate by replacing his break-bulk cargo ships built in the late fifties with new unitized cargo vessels, he had the additional problem of disposing of his current fleet.

Mandatory Capital Reserve Deposits

During the sixties, the subsidized lines were required to deposit annually in their capital reserve funds an amount equal to the annual depreciation on their fleets. These funds could be used only for vessel construction, conversions or repairs, or for containers or lighters. Edward Aptaker, Chief, Office of Government Aid, explained this policy: "Our interest is in having as much capital in the American flag steamship industry as we can reasonably keep locked into it."[10] Without such restrictions, the unsubsidized lines could spend their cash flow from vessel depreciation on such operations as terminal facilities, or even use this in nonmaritime investments.

Other Corporate Restrictions

During the sixties, the subsidized lines were quite restricted in their spending of funds outside their vessel operation. They could not withdraw annually as profits more than 10 percent of their capital necessarily employed. In addition they were generally prohibited from using this money to employ affiliates to supply ancillary shipping services, such as stevedoring, ship repairing, or towboat services. Moreover, if a subsidized line earned profits over 10 percent of the capital necessarily employed over a ten-year period, the subsidy recapture provisions required that half of these excess profits be returned to the government with the remaining half being deposited in the line's special reserve funds.

Since the Maritime Administration had to calculate the capital necessarily employed for each subsidized line, it considered each expenditure made by the company. A subsidized firm had to ask for approval from the government if it wanted to pay an executive over $25,000 a year, buy a company car, or purchase any other business asset not involved in the ordinary course of business in order to have these expenditures regarded as "capital necessarily employed." These procedures resulted in abnormally large amounts of bookkeeping for the subsidized lines.

Since the subsidized lines were impeded in the return on investment they could earn, this could have discouraged them for investing in high risk–high return innovations. In addition, outside investors might have preferred to place their money with companies whose return on investment was not hindered by government regulation. Note that for the years 1965–68, no subsidized line listed in *Fortune's* "Fifty Largest Transportation Companies" surpassed the return on investment produced by Matson and Sea-Land, as shown in Table 7-5.[d] Since the subsidized and unsubsidized lines operated in different environments, the values of R.O.I. in Table 7-5 are not directly comparable. However, this exhibit does show that by innovating Sea-Land and Matson

[d]Smaller companies not listed in *Fortune's* "Fifty Largest Transportation Companies" might have had a greater R.O.I. than Matson or Sea-Land. In the future, however, smaller companies will not be able to take advantage of economies of scale possible with unitized cargo.

Table 7-5
Steamship Lines in *Fortune's* "Fifty Largest Transportation Companies"*

Steamship Line	*Net Income as a Percentage of Invested Capital* 1965	1966	1967	1968
McLean Industries (Sea-Land)†	21.4	18.3	20.5	20.3
Matson Navigation	11.3	13.4	9.6	9.9††
United States Lines	6.7	5.9	0.02	–
American President Lines	9.5	10.2	9.6	8.5
Moore-McCormack	1.9	8.3	——	——
Lykes	N.A.	N.A.	3.6	6.9
American Export Isbrandtsen	8.0	5.7	2.3	4.9

N.A. - Not Available. This company not listed in *Fortune* that year.

– Indicates a net loss for the company that year.

* *Fortune* July 15, 1966, pp. 258-259, June 15, 1967, pp. 222-223, June 15, 1968, pp. 214-215, May 15, 1969, pp. 196-197. The fifty companies chosen are those with the highest operating revenues.

†The vast majority of the business of McLean Industries was conducted by its subsidiary Sea-Land. In 1969, McLean Industries merged with R. J. Reynolds Tobacco Company, whose transactions do not appear in this exhibit.

††Matson was merged into Alexander and Baldwin in March 1969. This figure is for Alexander and Baldwin, including Matson Navigation.

have been able to get a higher return on their investments operating in their unsubsidized environment than their corresponding subsidized counterparts.

Recommendations for the Subsidy Program in the Late Sixties

The best-made plans of subsidized lines can fail if government regulations hinder their implementation. Therefore, it is important that subsidy laws devised in 1936 for break-bulk cargo operators be revised for the unitized-cargo era. Specific changes in government regulations must be made to improve the ability of the subsidized lines to (1) build ships with a minimum of delay, and (2) operate their businesses with a greater degree of flexibility.

Subsidized Building Program

The CDS program is a subsidy only to the U.S. shipyards and various U.S. manufacturers and by its bureaucratic nature has delayed the building programs of the subsidized lines. In the late sixties, the two following methods provided a means to eliminate or greatly reduce these delays:

1. Allowing the subsidized lines to build abroad without being penalized.
2. Starting a massive shipbuilding program so that the subsidized lines would not have to wait for CDS funds and the U.S. shipyards could make the necessary capital outlays for equipment to speed up their building process.

The first solution had been politically unfeasible. Although the second solution had been proposed many times, congressional appropriation of funds for a large-scale building program had not been forthcoming.

Four less radical ways of reducing the delays in subsidized construction were as follows:

1. Granting of permission to convert subsidized ships in foreign yards. This would hasten the conversion of ships and leave more CDS funds for new construction.
2. Allowing new subsidized construction to be built in foreign yards after a certain amount of CDS funds had been allocated during one year.
3. Allowing subsidized lines to build new ships in foreign yards after they had waited a year for appropriation of CDS funds after having their CDS applications approved by the Maritime Subsidy Board.
4. Allowing subsidized lines to negotiate directly with the shipyards. In this way, a subsidized firm has the option of paying a higher price for earlier delivery.

Some of these alternatives have been suggested in the past without success. The Interagency Maritime Task Force recommended that subsidized lines be allowed to construct their ships in foreign yards after approximately $150 million in CDS had been allocated.[11] Although CDS funds have traditionally provided a small percentage of the revenue received by U.S. shipyards, the political power of the shipyards' managements and their workers have been able to defeat any such proposal. The shipyards feared that permitting the building of any U.S. merchant ships overseas could set a precedent to be followed for U.S. Navy ships, whose construction make up the majority of the shipyards' revenues.

During the sixties the subsidized lines had continually campaigned for more freedom in obtaining new vessels, but with little success. The ability to negotiate directly with the shipyards for their contracts would have been particularly helpful to the subsidized lines while they were eager to obtain new unitized-cargo vessels to meet the competition. Since the government would determine the CDS allocation for normal building conditions, any additional funds required for quicker delivery could have been borne by the steamship company without the danger of increasing the subsidy expenditures.

A more conservative alternative would have been to eliminate delays caused by a lack of congressionally appropriated funds by providing loans guaranteed by the government to obtain the necessary money until the next congressional appropriation. The question of who should pay the interest on the loan would have to be answered. One solution would be to divide the interest equally between the subsidized operator, the shipyard, and the government; however, each party probably feels that the interest should be divided between the other two.

Planning and Operational Flexibility

The unsubsidized lines have performed a public service by bringing innovation to the maritime world, and any government program should not discriminate against them. Nevertheless, archaic government regulations must be updated. In the late sixties, the following changes in government regulations pertaining to the operations of the subsidized lines would have greatly aided them:

- Consolidation of the trade route structure into eight to twelve trade regions, allowing each subsidized operator currently operating on a trade within each region to have free movement within the region as long as it meets its minimum and maximum sailing requirements.

Five provisions recommended by the author that were included in the Merchant Marine Act of 1970 were:

- Repeal of the requirement that subsidized vessels must be seventeen years old in order to be sold to the government upon buying new construction.
- Abolishment of mandatory Capital Reserve Fund deposits.
- Repeal of the provision prohibiting operators from withdrawing annually as profits more than 10 percent of their capital necessarily employed.
- Repeal of the provision prohibiting subsidized operators from generally employing affiliates to supply ancillary shipping services.
- Repeal of the operating differential subsidy recapture provision.

Many of these recommendations had been previously proposed by others. The subsidized lines had continually campaigned for any change in government regulation that would give them more control over running their businesses.

All these suggestions would aid the subsidized lines by eliminating much of the bookkeeping done solely to meet regulations and by giving subsidized companies more freedom in their operations. While the lines often obtained permission for changes in their route structures and their sailing requirements, applying for changes was time consuming, and competitors sometimes tied up such procedures in lengthy legal maneuvers. The legislation gives the subsidized lines more freedom in their vessel operations without altering their basic liner form of service or penalizing their unsubsidized competitors. Government safeguards are maintained by periodic auditing of the companies' books as is done for normal businesses.

The consolidation of trade routes, not yet allowed by law, would give the firms the power to expand their operations to take advantage of economies of scale in unitized cargo operations. With more flexibility in choosing trade areas, liner firms would have more opportunity to choose better inland and port locations for cargo consolidation. Some persons may feel that allowing subsidized lines to expand their operations over more trade routes would only lead to destructive rate cutting and disaster. However, chaos does not exist even though over 70 percent of the U.S. commercial liner commerce is carried by foreign-flag carriers. Since the majority of the vessels serving the United States that are foreign are not severely hindered by U.S. government regulations, there is no reason to penalize the small segment of the ships that are subsidized.

The legislation to allow subsidized lines more freedom in the management of their finances has provided the companies the opportunity to operate in a normal business-like manner in a competitive environment. By eliminating the restrictions locking funds into the steamship industry with mandatory Capital Reserve Fund deposits, the managements of the subsidized companies now have the opportunity to earn a better return on investment for their stockholders. Sea-Land, which has built more than $100 million worth of vessels, has shown that it is not necessary to have mandatory reserve fund deposits in order to build ships.

The new government regulations eliminate the need for the Maritime Administration to calculate the necessarily employed capital for each subsidized line. Consequently, the system of subsidy accounting is discontinued and the subsidized companies no longer have to ask approval for expenditures for their business operations. These changes will save the subsidized industry an enormous amount of bookkeeping, and the taxpayers the cost of the government machinery involved.

Effect of the Merchant Marine Act of 1970

The Merchant Marine Act of 1970 provided the subsidized liner firms with additional planning and operating efficiency. The subsidized lines were no longer bothered by restrictions relating to the seventeen-year old requirement on trade-in vessels, mandatory Capital Reserve Fund deposits, provisions restricting the annual withdrawal of profits, provisions prohibiting the employing of affiliates to supply ancillary services, or the recapture provisions. While the new legislation provided help to the subsidized liner firms in many areas, the author again recommends a broadening of the trade route regions, which would also result in more flexibility in sailing schedules.

The Act calls for CDS funds of $300 million per year for the next ten years and allows lines to negotiate directly with the shipyards under certain circumstances. Initially, it appeared that the supply of CDS funds would exceed the demand, even though the new legislation allowed the CDS funds to be spent on bulk carriers as well as liners. This result would have eliminated previous building delays incurred by liner operators caused by a lack of CDS funds.

Unfortunately, by October 1973 there were more than one hundred vessels with CDS applications pending. While only one liner vessel was included on the new construction waiting list, it was not obvious how many years this subsidized liner operator would have to wait to obtain CDS funds. In addition, at least two subsidized liner operators, desired CDS funds to convert break-bulk cargo ships to tankers. Meanwhile the Maritime Administration had not announced what criteria were being used to allocate CDS funds. The author feels that the recommendations of the late sixties pertaining to the elimination of building delays are still appropriate, so that subsidized liner operators will not get bogged down by bureaucratic regulations in obtaining new vessels.

Risks and Benefits

Each time a change is made providing more freedom for subsidized management to improve their operations, the danger is also present for poor management decisions to cause economic disaster. Increased flexibility could result in the failure of a subsidized line from either a poor investment in new vessels or poor business judgment outside the maritime field. In either case the major problem concerning the Maritime Administration would be the disposal of the company's vessels. If other subsidized lines did not wish to purchase them, the government could buy them and either charter them for commercial or military use or place them in the government reserve fleet. At the end of fiscal year 1969, the 1,050 vessels owned by the government were all over twenty years of age. By the end of fiscal year 1971, only 145 general dry cargo ships are expected to be available for emergency use.[12] Any operational ship would be a welcome addition to the government reserve fleet.

At the same time, the subsidized lines that were able to take advantage of their new freedom in operations and planning would grow stronger and produce a better U.S. merchant marine fleet. Although the proposed recommendations present the risk of economic disaster for some, the author feels that the benefits which would accrue to subsidized lines with progressive managements and to the U.S. merchant marine as a whole far outweigh these risks.

This chapter has looked at the subsidized and unsubsidized firms in terms of attitude, commercial environment, government regulation, and planning analysis. It takes a favorable combination of these factors to bring about success. The Merchant Marine Act of 1970 gives the subsidized lines more freedom in the planning of their resources; consequently, this places still more importance on their planning ability. Chapter 8 presents a regimen for planning subsidized ship replacement taking into account all these factors.

8 Proposed Methodology for Planning Subsidized Vessel Replacement

The decision on vessel replacement is an extremely important part of long-range planning, influenced by many interdependent factors which can vary to form an infinite number of combinations. The major problem in the industry's planning process has not been recognizing the factors to be considered, such as cargo demand, vessel types, labor unions, government regulations, inland transportation and port facilities, Rather, gathering the data and analyzing it in a systematic manner have been the weak points in planning. In the first place, steamship companies have rarely assigned personnel solely to planning ship replacement. In the second place, innovations, in the technology have been few, and planners pressed for time could safely make sweeping assumptions about the markets, competitors, port systems, and so forth, based on their current status. With unitization, such assumptions are inadequate for planning purposes.

Moreover, new forces are at work which must be accounted for in the planning process. For example, because containers facilitate the transfer of cargo from one transportation mode to another, inland networks of transportation may be used to extend an ocean carrier's area of influence. Accomplishing this requires incorporation of data on the ultimate origin and destination of cargo with the inland rates and practices.

The Planning Process

Economies of scale and other factors relevant to long-range planning will all be analyzed in a format outlined in Figure 8-1. Starting with the present and forecasted market demand, the company must look at how this cargo could best be carried if there were no such current restrictions as lack of port facilities or inland transportation systems, labor rules, customs procedures, etc. In this way the firm can develop some idea of what the industry may be like in five years, ten years, or twenty years from now as it continues to rationalize its operations. Next, a company must look at what is possible in the present environment and analyze its own strengths and weaknesses as well as those of its competitors.

At this point, company executives will have to formulate the alternatives open to their firm and to test the risks and advantages of each. They must then decide what role the company can play most effectively in the future

Define System

Cargo Market Analysis and Forecast

↓

Find Theoretically Ideal Solution

↓

What Is Possible Today

↓

What Is Being Done Today

↓

What Is Planned by Competitors

↓

Self Analysis of Strengths and Weaknesses
Comparison With Competitors

↓

Long Range Strategy and Vessel Design

↓

Obtaining CDS Funds

↓

Updating the Analysis

Note: The arrows in the diagram do not mean that the process cannot be iterative.

Figure 8-1. Proposed Strategy.

environment, what types of service it should offer five years, or ten or twenty years from now. Previous steps in the process will have included presumptions on the economic and physical characteristics of the company's fleet, but only after a company has developed a corporate strategy and long-range planning program are the details of implementation decided upon. The ships must be selected to meet the company's objectives.

Subsidized lines require CDS funds to build new vessels. In order to obtain construction differential funds for its proposed ships, modifications in vessel design may be necessary. No analysis is so accurate that it will serve long without revision. The final feature of the planning process will establish machinery for updating the analysis and altering strategy or ship designs if necessary.

The proposed planning process will require trying many alternative solutions to various problems. In addition, sensitivity analysis must be performed to identify the more critical factors. Although not essential, computer simulation techniques would be very valuable throughout this procedure.

Market Demand

A key factor in the proposed strategy is the forecast of market demand. With the advent of unitization, the problem becomes more complex and more critical. Since ships will be more specialized in the cargo they handle, it is more important to analyze specific segments of the market rather than the aggregate general cargo market. Because the market is being fragmented, the problem of categorizing the cargo and predicting the growth for each segment becomes more complicated. Unless the data are properly structured, management will be limited in the alternatives it can analyze.

The present and future demand for cargo must be analyzed not only on the trade routes served by a company but also in areas which could be affected by diversion of cargo. For example, with the use of unit trains or waterborne feeder service, cargo now in adjoining areas not served by a company could be diverted to those ports presently served or predicted to be served. A subsidized firm may also want to consider the possibility of sailing on new trade routes. These considerations will determine the regions in which the company should analyze trade.

Once a subsidized line has defined the geographic boundaries of its chosen system, it must look at the total flow of goods within this area. The analysis must even include cargo which the company believes it has no intention of carrying, because only by viewing the total system can the firm be confident of seeing all relevant factors. For example, a liner firm may not be interested in carrying shipload lots of grain eastbound in its trade, preferring to leave this bulk cargo for tramp operators. However, by considering the volume and types of ships moving eastbound the company can better predict the freight-rate competition between liners and tramps on the west-bound voyage.

Origin and destination data on cargo movements, rather than just port-to-port information, are desirable whenever possible. Such knowledge will aid in consolidation of freight.

The demand for cargo must be broken down into: (1) bulk cargo—wet and dry, (2) general cargo—containerizable and noncontainerizable, and (3) roll-on/roll-off cargo. Special commodities, such as coffee, which make up a significant segment of the cargo on certain trade routes, will also deserve special attention. Some of the cargo mentioned above may belong in more than one category. For example, vehicular traffic, although belonging to the roll-on/roll-off category, might also be containerizable. This should be noted in the analysis. All cargo must be analyzed in relation to density and freight rate per revenue ton. Cargo must be divided between commercial and government cargo since the profit on each may be quite different. This data will show the volume required by the cargo as well as its profitability.

Categorizing cargo types and forecasting demand is an extremely difficult and expensive task for a steamship line. However, once the information has been gathered, a company can use it in its operations and scheduling as well as in its long-range planning. Hopefully the problem of gathering data will be lessened by research financed by the government. The Department of Commerce has sponsored studies on projections of bulk cargoes.[d] In addition the Department of Transportation has sponsored studies which include forecasts for demand of containerizable goods.[e] The government also sponsored two studies, at a cost of about one-half million dollars each, aimed at determining ship designs to meet the needs of the future.[f] These types of studies should be particularly valuable to companies looking at certain types of cargo or particular trade routes that are not within the firm's present sphere of operations.

Analyzing the Ideal System

Once the company has made its cargo forecast, it should consider the most economical means of providing the origin-to-destination movement of all the cargo within this system without regard to such restrictions as current labor practices, government regulations, port facilities, lack of inland transportation or intermodal cooperation, technology, or other generally practiced methods of operation. In considering the origin-to-destination movement, a firm must analyze the costs of packaging and handling at the factory as well as at the port. The cost of inland transportation and freight consolidation must also be considered.

By disregarding current restrictions, a company can get an idea of the size and type of fleet along with the inland transportation networks needed to carry all the cargo in the system under ideal conditions. This may include vessel types and feeder systems not in use today.

In this stage of the analysis, ships need only be defined by the following parameters: construction cost, operating cost, carrying capacity, speed, and cargo-handling rate. Note that since the economic analysis places a value on time as well as transportation cost, air freight will be a competitive factor in these studies.

---

[d]Such studies are the *Forecast of U.S. Oceanborne Foreign Trade in Dry Bulk Commodities* Booz-Allen Applied Research, Inc., Bethesda, Md., 1969, and *Projections of Principle U.S. Dry Bulk Commodity Seaborne Imports and Exports for 1975 and 1995,* Stanford Research Institute, Menlo Park, Calif., 1969.

[e]These studies are *Oceanborne Shipping: Demand and Technology Forecast* by Litton Systems, Culver City, Calif., 1968, and *Transoceanic Cargo Study,* by Planning Research Corp., Los Angeles, Calif., 1971.

[f]Newport News Shipbuilding and Dry Dock Co. and Bath Iron Works Corp. were the two prime contractors for these studies.

Naturally a study of this sort cannot expect to be completely accurate. However, general trends can be predicted and untried methods of transportation can be tested at least tentatively. This section of the analysis invites the company to look beyond presently practiced methods, thus providing the opportunity for innovation. Men who were not constrained by the status quo have been responsible for 33 knot containerships, and barge-carrying vessels. By looking at more than just one segment of the cargo market, Matson developed combination container-bulk sugar ships and combination container-automobile carrying vessels.

Determining What Is Possible Today

Once the company has considered what could happen under ideal conditions, it should return to reality. This step should distinguish between real obstacles and those created by a failure to rationalize the system. Upon examination, the line may discover that certain vessel types or feeder systems are in fact practical even though unused today. A long distance container train currently not in use because of lack of cooperation between railroad companies or certain countries might be arranged. Consolidation of cargo at inland points might be economical by building certain facilities. Certain capital-intensive operations not currently in use may need only new methods of financing to become possible. Here is the occasion to improve the status quo.

Of course, the company may also discover that many operations it had planned are currently not feasible. Maybe inland transportation of an underdeveloped country cannot support a proposed containership operation. Possibly canals are not wide enough to take certain lighters the company had in mind. Port congestion might make necessary turnaround times impossible. Customs regulations between countries might hinder the planned inland movement of cargo. Structural difficulties may prevent the construction of ships as large as planned. Companies using private or contract carriers in certain areas might make consolidation of particular cargoes impractical.

Artificial cost barriers may also bar the most economical movement of cargo. For example, longshoremen rates might favor lighters over ships. Seamen's wage rates may favor using a tug and barge combination rather than a self-propelled ship. Government restrictions giving cargo preference on certain goods to ships of its own flag might make certain backhaul assumptions in the ideal situation impractical.

All of these factors that prohibit the ideal solution will modify the size and type of fleet needed. This fleet structure must be recalculated to find the best way to move the cargo under present conditions. Note that the fleets of individual companies have not yet entered into the analysis. Except for cargo which is required to go in bottoms of certain national flags, the remaining goods simply travel in ships unspecified as to company or country. However, at this point the company can realize how current restrictions are hindering the ideal solution. The line can also calculate the economic effect of removing certain obstacles, either technological or institutional.

The Present Environment

Now the company must look at how the cargo in the system is actually being carried today. The firm will probably discover that many of today's inefficient ships would be forced out of business if the ships proposed in the previous analyses were built. Inefficient lines currently operating may be totally dependent on the artificial rate structure of the conference system for their existence. The fleet providing the cargo movement today is a result of many factors in the environment. The factors below all affect the movement of cargo; therefore, they all play a part in determining the size, speed, and type of vessels which are currently used.

Warbuilt Fleet. Many ships currently in service were built during World War II. The purchase of these ships at artificially low prices had discouraged the construction of new vessels of different designs. However, as these vessels become economically obsolete in the next decade, they will be replaced. The new fleet will probably be more in line with the outcome of the previous analyses examining the most economical movement of cargo.

Port Conditions. The various ports in the system may differ in terms of depth of water, terminal facilities, amount of port congestion, safety from storms, and distance from the ocean. To reach some ports may require passing through locks to reach a particular harbor, incurring a time delay.

Labor Conditions. The labor conditions at various ports as well as aboard ship may vary by union or region. Ports anxious to obtain a share of the container business might be more liberal in their stevedoring terms. The cooperation of various seamen's unions will determine which company or which country can use certain advances in ship automation.

Government Regulations. Regulations by various government agencies relating to customs procedures, bills of lading, etc. affect cargo movement. In addition, regulations for carrying cargo between countries inland may influence the type of transportation chosen. Subsidized lines are affected by gaining permission for changing foreign areas served and frequency of sailings. Cargo preference offered by many governments to ships of their own flag is reflected in present fleet operations.

Inland Operations. The inland operations serving a port are important, whether they are inland waterways, truck or rail. The possibilities of inland container pools and container trains will affect the ability to consolidate large amounts of cargo needed for huge unitized cargo ships.

Financial Considerations. Entering the unitized cargo market will entail a considerable investment in ships, containers, terminal facilities, and loading and unloading equipment. In addition, start-up costs will be incurred in establishing a computerized container system, retraining salesmen, and setting up a research staff.

Construction Differential Subsidy Funds. Since U.S. subsidized lines require CDS funds in their ship construction, the availability of such funds will affect the amount of new subsidized construction. Standardized vessel designs may appear as a result of government pressure in trying to find ways of lowering construction costs. A company may also decide that rather than wait for new construction or CDS funds it would be more advantageous to convert its present ships or buy ships from another company.

Conference Agreements. The use of conference and various pooling agreements may affect the design of vessels. If a line cannot charge lower prices to attract more cargo for a large economical ship, there is less incentive to take advantage of economies of scale in the size of vessels. Also if a company cannot compete on price, there is an incentive to build ships faster than would otherwise be economically optimal. In this way, a firm can have an advantage by providing faster service at the same price.

The Competitive Environment

By now the company has investigated how the cargo moves within its system and has realized the obstacles to improving the system. At this point, the company must analyze its strengths and weaknesses including share of market, types of cargo now carried, number of vessels as well as type and age, other equipment, management, sales force, trade routes, subsidiaries, and finances. This appraisal must also be duplicated for its competitors including their vessels under construction and their plans for the future. Then the company can evaluate who is now providing these services needed for the future.

Another factor to be considered is the action/reaction effect of one firm to the change in the operations of its competitor. For example, consider a situation in which one company places a full containership in the liner market of a system which is composed entirely of break-bulk cargo operators. A number of "what if" simulations will show the effect on competitors' operations of various numbers of full containerships using various lower freight rates.

Note that a company must study the cost structure of its competitors as well as its own in order to understand the process of action and reaction in the environment. Under certain market conditions, a subsidized line may feel that changing from break-bulk cargo ships to full containerships will produce only a marginal operation. However, an unsubsidized line looking at the exact same situation may view it differently. Since the unsubsidized line does not receive up to $700,000 per ship in ODS funds, it realizes that it cannot compete side by side against a subsidized operator in break-bulk cargo operations. By using a capital intensive container operation, the unsubsidized operator can reduce the competitive advantage of ODS funds. For example, if an equal amount of cargo per year can be moved by the containership operation of an unsubsidized line with only one third as many ships as its subsidized competitor using break-bulk cargo vessels, the competitive advantage of ODS funds has been reduced two thirds.

A subsidized break-bulk cargo operator must also realize that the economies of scale with unitized cargo operations can work to his disadvantage. For example, a larger established operator of containerships might be able to start service on a new trade by adding only one container terminal operation and one sales office, and by rescheduling a few ships. Not only does this allow a low start-up cost for the containership operator, but the added cargo obtained on this trade route may be handled at very little incremental cost in the high fixed-cost container operation. Simulations for various "what if" situations may show the subsidized break-bulk cargo line that despite its reluctance to containerize, containerships will eventually come to its trade route and that its only choice will be to start a unitized cargo operation or be driven out of the liner business.

At this point, the subsidized line knows where it stands relative to its competitors. Equally important, it knows how the present methods of transportation can be improved in providing the origin to destination movement of cargo in its system. Because the company is aware of the optimum system and changes in the present system, it has an idea what the fleet in its system will look like five, ten, or twenty years from now.

Strategic Planning

With this information, each subsidized line must decide what role it wishes to play in the future. A company may have to decide whether it is willing to make a major change in its operations. A break-bulk cargo operator, for example, must decide if it wants to become a large-scale carrier of unitized cargo, a change which will require a huge investment and equipment.

Just as there is the danger that a company may blindly refuse to change its mode of operation, there is also the danger that a company may blindly change its mode of operation without properly examining its environment.

As was noted in each of the case studies, one alternative open to a subsidized line is to retire from the steamship business. In fact, the influence of profit-oriented conglomerates in the steamship industry may cause the elimination of some steamship operations. Another alternative is to become an unsubsidized line. Because containerization is subject to economies of scale, a small liner firm may elect to merge with another steamship company. The analytical steps up to now will act as a valuable tool in determining the firm's long-range strategy; however, a program for planning is only as creative as the alternatives posed and only as accurate as the management that evaluates it.

Long-Range Planning and Vessel Design

Assuming the company has decided to continue as a subsidized operator, the most important factor necessary in carrying out the long-range strategy of the company will be its choice of vessels. Most subsidized liners currently are in the process of building or planning new construction or conversions. These activities are a result of a desire to adopt new technology or to fulfill a contractual obligation to replace their vessels. With the company's long-range strategy determined, the data already gathered can be used to provide the choice of ship design necessary to meet this objective.

Just as choosing the ship design is part of long-range planning, there are many other related factors involved in a decision to become a unitized cargo carrier. "What if" simulations using the data from previous analyses will be a valuable management aid in making many of the following decisions: the number and sizes of containers or barges, whether to buy or lease them, whether to establish inland container pools, and whether to own and operate marine terminal facilities. New personnel will probably be needed, and the sales force will need retraining. A marketing strategy must be developed to meet the needs of the shipper rather than simply allowing him to use the port-to-port service provided by the company.

Obtaining CDS Funds

If a subsidized line has decided to change from a break-bulk cargo operator to a carrier of unitized cargo, it is imperative that it build or convert ships to carry out this strategy. The best planning process can result in dismal failure if the Maritime Subsidy Board will not allocate CDS to a firm for new construction. Since the board has favored standardized ship designs for more than one company, a subsidized line must look at the consequences of choosing an established ship design. If the company has created a new design, it should consider trying to convince other companies to use this design too. Since fiscal year 1967, the board has stressed productivity in its CDS allocations for liner vessels. Consequently, a company must consider whether the cost of making its design more productive, larger and/or faster, is worth the advantages, if any, in helping it to receive CDS funds.

If the CDS funds appropriated by Congress are not sufficient to finance the proposed new construction of the company, the firm may be forced to settle for converting present tonnage, a move which would require a smaller CDS allocation. Converting present tonnage has the advantage of being quicker as well as requiring a lower investment. Naturally, conversions have the disadvantages that part of the company's fleet is temporarily out of use, the resultant ship design probably is less than ideal, and its life is shorter than that of a new ship.

Updating of the Analysis

Since a company may wait eight years from the time its original initial decision is made until the final ship in that flight is in service, a subsidized line must be able to evaluate changes in the environment and, if necessary, be prepared to change its design and even its long-term strategy. Sensitivity analysis will show that the choice of ship design will probably be based on a number of critical factors, such as a high growth rate for containerizable cargo, liberal work rules for unloading cargo from barges, or a certain construction cost for a new type of ship.

At the time a decision on ship design is reached, procedures should be established to continue measuring these critical factors. If a critical factor changes unfavorably, an alternative strategy must be available. Periodically, perhaps every six months, the new information on these critical factors should be recorded and used to update the original analysis. Such factors as unexpected growth or technological breakthroughs in construction techniques should be analyzed to see how they affect the initial decision. After ship construction is completed, the company should continue updating the analysis so that it can be used in the long-range planning process between vessel replacements. The analyses may also prove valuable in determining schedule or route changes, effects of new labor rules, and effects of actions by the competition.

Organizational Structure

The proposed strategy for planning vessel replacement will require more time and effort than most subsidized lines are currently allotting to this function.

However, when a company is considering an investment of over $100 million, the preinvestment analysis is not the place to cut corners. Since the penalty for designing the wrong ship can be the failure of the company, the importance of proper analysis should not be underestimated.

The number of staff members assigned to research and planning in the past has varied widely with different companies. Some subsidized lines have relied on executive committees to analyze data collected by operating personnel. On the other hand, Matson managed to build up a staff of more than thirty persons devoting themselves fully to the task of research; however, this group carried outside contracts in addition to serving the internal needs of Matson.[d] Most subsidized lines will probably decide to organize somewhere between these two extremes.

Every company may now need a corporate planning director who has both the responsibility and authority associated with long-range planning and corporate strategy. At the very least, this man can keep up to date on published research and outline and review work done by outside consultants. Hopefully, he can organize a small staff inside the subsidized line which can devote its energies to researching and analyzing the environment for planning purposes.

The exact organizational structure to cope with the planning process will vary according to the environment and long-range strategy of each company. A company whose strategy is to become a worldwide supplier of integrated transportation services and logistics can expect to build up a substantial research staff. On the other hand, a very small break-bulk cargo operator who services only a small share of the market and desires to continue in this strategy, may feel that he need not put much emphasis on planning. However, it is imperative that this company analyze the suitability of container operations of its trade route and realize the consequences of having full containerships as competition.

Many companies may feel that because of lack of internal competence in certain areas, outside consultants should be used to a large extent in the planning process. This may be appropriate for some companies; however, the subsidized line must have the capability to understand and evaluate a consultant's report and the expertise to implement the course of action recommended.

The management of a steamship line providing break-bulk cargo service for several decades will now have to acquire competence in the area of intermodal transportation in order to enter the unitized cargo market. This knowledge is not so critical for companies operating barge-carrying ships as for containership operators. In some instances the barges can be emptied and loaded directly at the port; thus the liner firm is providing only a port-to-port service. However, to the extent that the barges carry general cargo

[d]This group has diminished in size to concentrate solely on the problems of Matson.

rather than just bulk cargo, a containership operator will eventually compete for this cargo, offering a door-to-door service. Consequently, to understand and evaluate his competition and also his general cargo market, the barge-carrying ship operator must understand the inland transportation system.

Probability of Success

The proposed strategy presents a process that forces management to look beyond the status quo and allows them to plan for the future in a systematic manner. Using this procedure will ensure that vessel replacement will become part of long-range planning.

It is not possible to pinpoint how much money should be spent on analysis. If a company is considering an investment of $100 million in new ships, containers, other equipment, and start-up costs, it is difficult to specify whether the funds provided for planning and analysis should amount to 5 percent, 1 percent, or one tenth of 1 percent of the total investment. However, if subsidized lines merely use planning processes of the break-bulk cargo era, even this last figure will not be met in many cases. Although a specific amount of money for future planning cannot be given, there can be no doubt that more emphasis must be placed in this area in the era of unitized cargo.

Using the proposed strategy for planning vessel replacement does not guarantee success. In the first place, it is not obvious that all necessary data of the origin-to-destination movement of cargo can be accumulated at present by expending a great amount of time and effort. Secondly, factors beyond the control of the subsidized line, such as government regulations, labor union rules, and erratic growth or decline of cargo movements of certain commodities, cannot always be accurately predicted.

Nevertheless, the strategy proposed here should increase the probability of success over a company that ignores formulating a long-range strategy or treats the problem superficially. Costly mistakes can be avoided if a company understands its environment. Current diseconomies in a system can be eliminated if a firm has done more than follow tradition. Changes in markets, competitors, strategies, etc. can be anticipated, and response to unexpected change can even be allowed for, if a company looks ahead imaginatively. Planning cannot guarantee success, but having no planning may guarantee incurring economic penalties. Companies that cannot afford to plan in the current complex environment of the maritime industry may not be in business tomorrow.

9 Summary and Conclusions

The Merchant Marine Act of 1970 calls for multibillion dollar public investment in the Merchant Marine over a ten-year period to ensure the construction of 300 new ships. Some of these funds will produce new liner vessels. Whether these funds will provide a modern competitive subsidized liner fleet or will be wasted on inefficient ships depends to a large degree on two factors: (1) the ability of the subsidized liner segment to improve its planning processes so that it can better discern market opportunities and choose fleets to meet them, and (2) the elimination of some of the present hindrances of government programs in providing construction differential subsidy funds and in regulating the general operations of the subsidized lines.

The various general assertions that are made do not necessarily apply to a particular line or to all liner firms at a particular time. In order to better understand the changes brought about by unitization, it is useful to start with a brief summary of the subsidized planning process during the break-bulk cargo era.

Subsidized Planning Process for Vessel Replacement During the Break-Bulk Cargo Era

During the break-bulk era, all liner firms used the same types of vessels and the shippers frequently could not differentiate between liner services. Rate conferences and pooling agreements helped to maintain this system. Consequently, the subsidized lines had relatively stable market shares and little need for long-range planning.

In this environment, the companies rarely had permanent planning staffs for ship replacement. Instead the men normally employed in the day-to-day operations of the company were the ones to make decisions on the new ship designs.

Consequently they normally simplified the planning process by adopting a standard or traditional method which made great use of management judgment and pro forma voyage calculations used in their daily operations. The criteria to be met by these designs apparently were the ability to get CDS funds, higher productivity than the ship it replaced, speed and size at least as great as any ship in the company's fleet, and speed as fast or faster than vessels used by competition on the same trade route. The criteria could normally be met easily by adopting or modifying the few basic ship designs of the time.

These men concerned with the daily operations of the companies competently handled the ship replacement planning process during the break-bulk cargo era. However, the advent of unitization made the process so complex that these men had neither all the skills nor the time necessary to handle the planning program.

The Managerial Implications of Unitization

The construction of containerships and barge-carrying ships have caused unitization to bring great changes to the maritime world. Unitization produced a capital intensive operation relative to the operation of break-bulk cargo ships. The investment of new specialized vessels was often equaled by the required investment in containers, bogies, and other equipment. Consequently, the financially weaker subsidized lines needed to obtain proper financing before starting a large-scale unitized cargo operation. This initial financial burden was compounded by the economics of scale possible with unitization in respect to operations of terminal facilities, computerized control of containers or barges, and the total required number of containers or barges. Therefore, a liner firm beginning unitized cargo operations on a small scale could be at a great cost disadvantage relative to a larger firm. Many smaller subsidized lines now must consider the alternatives of going out of the steamship business, becoming unsubsidized and providing tramp services, merging with maritime or nonmaritime companies, or forming a consortium with other liner firms.

Containerization marked the start of a door-to-door rather than a port-to-port service. Consequently, information on the origin and destination of goods, inland transportation, inland container pools, and container control system has now become important. Because inland carriers are generally more concerned with domestic traffic—which makes up the vast majority of their business—rather than international cargo movement, management of liner firms has become responsible in part for coordinating the ocean-borne traffic with inland traffic movements.

With the advent of unitization, huge specialized ships are now in service and under construction. This increase in vessel size had been inhibited with break-bulk cargo ships for two reasons: first, it was not worthwhile building giant ships so long as a greater part of their economic capability and capacity was lost by long periods spent in port; second, because of a large number of competitors using the same types of ships, building a giant vessel would not guarantee sufficient utilization. With containerships entering the U.S. foreign trade, the past relatively stable market shares in the liner field have been disrupted, and the number of competitors has already been reduced on certain competitive trade routes. With the new opportunity to take advantage of larger ships, management must now have a better knowledge of cargo growth and thus must study the segment of the cargo market that can be physically and economically carried by its fleet of specialized ships. With their operations becoming more capital intensive, companies must try to minimize the number of ports served. This leads to analysis of inland transportation, including feeder systems and container trains as well as how this transportation can be used to consolidate cargo and to divert it from other trade routes. While marketing in the break-bulk cargo era could consist of taking a customer to lunch, new marketing approaches are now necessary to analyze and meet the needs of the customers. It is evident that the advent of unitization requires the development of new skills, new knowledge, and new personnel in the subsidized lines.

Planning Processes for Subsidized Ship Replacement after 1961

The failure of the attempt of Grace Lines in 1960 to become the first subsidized line to provide full containership service helped to discourage other subsidized lines from altering their break-bulk cargo operations. The following reasons have been cited for the delay by subsidized lines in bringing large-scale containerization into the U.S. foreign trade: absence of specialized expensive equipment, expanding operations with nonstandard sizes and fittings, customs complications, archaic government regulations, steamship conference objections, lack of shipper-carrier enthusiasm, the long investment life of steamships, and the longer periods of time needed to build new subsidized ships in U.S. shipyards versus the construction time in foreign yards.

In 1959, the National Academy of Sciences completed a report which revealed the potentials of container vessels and the liabilities of semicontainerships.[1] The calculations showed that if the rate of loading and unloading containers could equal eight times the cargo-loading rate of break-bulk cargo it would be economically advantageous to build a full containership rather than a semicontainership. Combining containers with break-bulk cargo to get a semicontainership would increase the cost of the break-bulk ship with the costs of container ownership or leasing, handling, and storing. Since the port time of the ship would be determined by the amount of break-bulk cargo, the combination ship would be more costly without being more productive. Therefore, semicontainerships would be combining the disadvantages of both methods.

By the time the National Academy of Sciences published their next report in 1963 covering the total cost of origin-to-destination cargo movement, the necessary container cargo handling rate had been realized.[2] Consequently, the study excluded the economically inferior degrees of containerization and went on to show the system benefits of using full containerships rather than break-bulk cargo ships.

Even though these studies showed that full containerships were economically superior to both break-bulk cargo ships and semicontainerships, many subsidized lines decided to use semicontainerships when it became apparent that either Sea-Land, Matson, or foreign competitors would use full containerships in the U.S. foreign trade. All planning processes of the four subsidized lines studied in depth in this report showed room for improvement. The subsidized lines either did not coordinate their new vessel designs with a long-range planning program or chose a long-range objective which was not supported by adequate facts.

States Steamship Co., American Mail Line, Moore-McCormack Lines, and Executive Lines all originally felt that ships capable of carrying both break-bulk cargo and containers were the type to build, while the competition was preparing to use full containerships. States and American Mail Line apparently surmised incorrectly that they could continue indefinitely as break-bulk cargo operators. For States, this seems to have resulted from wishful thinking and poor analysis. Although AML made an analysis of its environment, it apparently failed to consider thoroughly the movement of cargo inland to and from the port or the economic consequences of competing against full containerships with semicontainerships. Moore-McCormack Lines and Executive Lines concentrated on the transition period between break-bulk cargo ships and full containerships rather than on long-range planning. Both companies overlooked the calculations showing that a semicontainership or a very flexible ship could economically compete with a full containership on the North Atlantic for only a very short period of time, if at all. Moore-McCormack later decided to sell its new ships and withdraw from the North Atlantic. Executive Lines decided to change its ship design twice, at an unnecessary additional cost of both time and money, so that it could better compete for container cargo.

Comparison with Sea-Land and Matson

Sea-Land and Matson, two unsubsidized lines, pioneered the large-scale use of containerization in the 1950s. Today both have grown and prospered in the maritime world.[a] Matson developed a containership fleet as a result of careful planning using a permanent planning staff with operations research capabilities. At each stage of the company's growth, various alternative courses of action were tested by a computer simulation that included the movements of competition as well as Matson's fleet. In this way the Matson management attempted to predict the outcome of using a large variety of input variables

[a]However, both have encountered their share of difficulties in recent years.

in their computer program. This method of planning helped Matson to construct unusual ship designs such as combination container-automobile carriers as well as more conventional full containerships.

In contrast, Sea-Land has now grown to control the largest privately owned full containership fleet in the world without using a formal planning staff. By utilizing the experience and intuition of its founder, Malcom McLean, who originally started in the trucking businsss, Sea-Land began container service first in the coastal trades, then in the noncontiguous domestic trade, and finally in the U.S. foreign trade. The management at Sea-Land was able to recognize its errors and respond quickly to correct them with only an informal planning program. However, Sea-Land has now set up a formal planning staff and plans to make greater use of computer simulation techniques in the future.

The environment leading to containerization was quite different for the unsubsidized and subsidized operators. After World War II, wages in the maritime field both aboard ship and on shore rose steadily. Since the domestic trade routes were shorter than the international ones, the labor costs made up a greater percentage of total costs for these voyages. In addition, the unsubsidized operators in the domestic trades did not receive ODS funds to pay most of their crew wages. Consequently, the domestic operators began to feel a profit squeeze. The intercoastal trade routes began to die under competition from the railroads whose labor wages had not risen as sharply. Even in the noncontiguous routes, operators had difficulty trying to raise the freight rates enough to keep up with the rise in labor charges. Caught with their backs against the wall, the unsubsidized lines dared to experiment with containerization.

A different attitude seemed to prevail in Matson and Sea-Land than in the subsidized lines. Matson and Sea-Land had a general atmosphere of desperation and personnel not entrenched in maritime tradition, whereas much of the management of the subsidized lines was made up of former merchant marine officers and relatives of founders of privately owned steamship lines. (It is human nature that the latter would resist changing to a different type of marine operation with which they were not familiar.)

The innovation of the unsubsidized lines was aided by their ability to obtain containerships at relatively low cost because they took advantage of the Ship Sales Act of 1946 and the Vessel Exchange Act of 1960. The first act allowed sales of warbuilt ships to both subsidized and unsubsidized operators. By the time the sales program closed in 1951, the subsidized lines had obligated themselves to use new construction to replace all their warbuilt tonnage in approximately twenty to twenty-five years after its original launching. The unsubsidized lines were not so obligated and proceeded to convert warbuilt tonnage to containerships at relatively low cost. The 1960 Vessel Exchange Act enabled the unsubsidized lines but not the subsidized ones to trade in vessels for better ships in the government reserve fleet. Consequently, Matson and Sea-Land were able to upgrade their fleets without paying for new construction. This gave them not only the advantage of low investment but also the advantage of the much shorter waiting periods for conversion than for new construction. Even when the unsubsidized lines decided to build fast, modern vessels, they possessed advantages over the subsidized lines. Since they were not subsidized, Matson and Sea-Land did not have to wait for government agencies to perform technological and economic studies on their proposed designs or for Congress to appropriate funds for construction in U.S. shipyards. Instead, Matson and Sea-Land were able to purchase their ships to be used in the U.S. foreign trade in overseas shipyards which had delivery time quicker than American yards.[b]

The flexibility available to the unsubsidized lines also aided the introduction of large-scale containerization. Unlike the subsidized lines, Matson and Sea-Land were free to change their schedules and trade routes at will, operate in both the domestic and foreign trades, charter U.S. or foreign vessels, obtain vessels from the government reserve fleet, and build in U.S. or foreign shipyards.

The overall planning process for subsidized vessel replacement is influenced by the attitudes of top management, the analysis used, the environment, and government regulation. It is hoped that the attitudes of top management of subsidized companies have become more progressive as conglomerates acquire these lines and as the profits of subsidized lines disappear. The remaining factor of the planning process over which the lines have the most control is market analysis. The studies of the four subsidized lines have shown that the subsidized industry can benefit from more emphasis on its analyses.

Proposed Methodology for Planning Subsidized Vessel Replacement

The thrust of this strategy is to combine long-range planning and vessel replacement in the same procedure. The new design must be created to implement a reasonable corporate strategy. By putting the proper time and effort on the initial analysis and then continually updating this analysis, the subsidized lines will be better able to produce vessels that will be economically successful for the twenty-five year life of the investment. The proposed regimen for planning consists of the eight-step procedure shown in Figure 9–1.

First a company must define its system in terms of geographic boundaries and types of cargoes considered. The boundaries on both categories are likely to extend beyond the company's present activities. Next the company must determine the theoretically "optimum" or lowest cost method of moving

[b]However, foreign-built ships registered in the United States were not allowed for three years to carry the 50 percent of government-sponsored nondefense cargo reserved for U.S.-flag vessels.

Define System

Cargo Market Analysis and Forecast

↓

Find Theoretically Ideal Solution

↓

What Is Possible Today

↓

What Is Being Done Today

What Is Planned by Competitors

↓

Self-Analysis of Strengths and Weaknesses

Comparison with Competitors

↓

Long-Range Strategy and Vessel Design

↓

Obtaining CDS Funds

↓

Updating the Analysis

Note: The arrows in the diagram do not mean that the process cannot be iterative.

Figure 9-1. Proposed Strategy

the cargo within the defined system without regard for such present restrictions as labor rules, customs laws, government regulations, or lack of intermodal cooperation. During this step and many following steps, a large number of alternative solutions will be tried. Thus computer simulation techniques will be very useful in this procedure.

After determining the theoretically "optimum" solution, the management must use its own judgment concerning what moves are possible at a particular time. (Hopefully, studies currently being financed by government agencies will be useful to lines in their own research.) At this point, the subsidized line can see how its present operations compare with what is being done by competitors, what is being planned, what is currently possible, and what is theoretically optimal. This overview will prove useful to the company in determining how changes in its environment will move current operations either closer or further away from the optimum.

Now that company must perform a candid self-analysis, including an evaluation of such resources as company fleet and equipment, management, staff, financing capabilities, and subsidiaries. This analysis must be repeated, as well as possible, for the company's competitors.

Assuming that the company wishes to continue as a subsidized operator, it will probably desire new construction either to meet the competition or fulfill its subsidy contract obligations. The new vessel design must enable the company to implement the long-range strategy it has picked. From the wealth of data already collected, management must choose a ship design that will maximize return on investment. A large number of simulations are useful here to avoid overlooking the best solution.

Once the design has been chosen, management must determine the best way to obtain CDS funds. If CDS funds are readily available, no problem exists. However, if such funds are in limited supply, subsidized lines must determine whether it is worthwhile to modify its design so that it will be better able to receive CDS from the Maritime Subsidy Board. From CDS allocations in the late sixties, it appeared that the board stressed productivity and standardized ship design.[c] Therefore, a company should consider the possibility of making its vessel design larger and/or faster. Also a line may want to copy an existing design, or convince another company to adopt its new design. Finally, the company must update its analysis until the new ships are in service in case factors in the environment change to make the original design undesirable.

The type of analysis described here should be a continuous process of gathering and evaluating data which can be used for planning routes and schedules as well as planning vessel replacement. The strategy proposed here is no magic panacea for the problems of the subsidized lines. However, this procedure will place emphasis on those factors that need improving. In addition, the subsidized lines will now be better able to understand their environment so that they can try to initiate favorable changes rather than preserving the status quo. Although instituting a formal planning process in the company will not guarantee success, if properly executed, it will improve the probability of success.

When a company enters an era of new technology with a $100 million investment, it should not cut corners on its preliminary analysis. A proper planning process will require a greater commitment of time and effort than most companies are now using. The organizational structure necessary to perform this analysis will vary with the needs of the company. Nevertheless, at least one person, and hopefully more, should devote his full time to corporate long-range planning; he must also have the responsibility and authority appropriate for the importance of this position.

Recommendations for the Subsidy Program

Government regulations can hinder and even prevent implementation of the best-made plans of the subsidized lines. Specific changes in government

[c]By the latter half of 1973, it was not obvious what criteria the Maritime Subsidy Board was using in the allocation of CDS funds.

regulations must be made to improve the ability of the subsidized lines to (1) build ships with a minimum of delay, and (2) operate their businesses with a greater degree of flexibility.

Subsidized Building Program

The CDS program is a subsidy only to the U.S. shipyards and various U.S. manufacturers and by its bureaucratic nature has delayed the building programs of the subsidized lines. These delays could be eliminated or greatly reduced by either of the two following methods:

1. Allowing the subsidized lines to build abroad without being penalized.
2. Starting a massive shipbuilding program so that the subsidized lines would not have to wait for CDS funds.

The first solution has been feasible politically in the past. Although a massive shipbuilding program was to have been brought about by the Merchant Marine Act of 1970, this has not eliminated building delays.

Three less radical ways of reducing the delays in subsidized construction are as follows:

1. Granting of permission to convert subsidized ships in foreign yards. This would hasten the conversion of ships and leave more CDS funds for new construction.
2. Allowing new subsidized construction to be built in foreign yards after a certain amount of CDS funds, such as $300 million, has been allocated during one year.
3. Allowing subsidized lines to build new ships in foreign yards after they had waited a year for appropriation of CDS funds after having their CDS application approved by the Maritime Subsidy Board.

A more conservative alternative would be to eliminate delays caused by a lack of congressionally appropriated funds by providing loans guaranteed by the government to obtain the necessary money until the next congressional appropriation. The question of who should pay the interest on the loan would have to be answered. One solution would be to divide the interest equally between the subsidized operator, the shipyard, and the government; however, each party probably feels that the interest should be divided between the other two.

Planning and Operational Flexibility

The unsubsidized lines have performed a public service by bringing innovation to the maritime world, and any government program should not discriminate against them. Nevertheless, archaic government regulations must be changed. The major problem regarding governmental regulations which many subsidized lines will face in the next few years will be the ability to obtain CDS funds for ship construction and conversion. Since more than one hundred vessels are on the list of CDS applications pending, it is difficult to predict when each subsidized liner operator could receive more CDS funds. The recommendations described above could alleviate the problem.

In order to help the operations of the subsidized lines, the author recommends the consolidation of the trade route structure into eight to twelve trade regions, allowing each subsidized operator currently operating on a trade route within each region to have free movement within the region as long as it meets its minimum and maximum sailing requirements. Currently, the lines can often obtain permission for changes in their route structures and their sailing requirements. However, applying for changes can be time consuming and competitors can sometimes tie up such procedures in lengthy legal maneuvers. The proposed trade route consolidation would give the subsidized lines more freedom in their vessel operations without altering their basic liner form of service.

The recommendations to allow subsidized lines more freedom in the management of their affairs would simply provide the opportunity for these companies to operate in a normal business-like manner in a competitive environment. As Manual Dias, former President of American Export Isbrandtsen Lines, stated:

> If the management of a subsidized liner company is so inefficient and so incapable of competing in today's business world, it should not be so protected—it should be allowed to go into bankruptcy.[3]

Since the granting of the above reforms would give the subsidized lines more freedom in the planning of their resources, this would place still more importance on their planning ability. With the forces of the marketplace more directly affecting the subsidized lines, the public investment in the U.S. Merchant Marine will produce a stronger subsidized liner fleet.

Notes

Chapter 2
Subsidized Environment before 1961

1. Alfred Murr, *Export/Import Traffic Management and Forwarding* (Cambridge, Md.: Cornell Maritime Press, 1967), p. 91.
2. Joseph G. Barkan, "Spotlight on the Construction Differential Subsidy Program." The Metropolitan Section of the Society of Naval Architects and Marine Engineers, March 19, 1963, p. 3. Published herein by permission of the aforementioned society.
3. *Ibid.*, p. 6. Published herein by the permission of the aforementioned Society.
4. Allen R. Ferguson et al., *Economic Value of the Merchant Marine,* Transportation Center at Northwestern University, Evanston, Ill., 1960, pp. 44-45.
5. Barkan, "Spotlight on the Construction Differential Subsidy Program," p. 7.
6. *Ibid.,* p. 8.
7. U.S. House of Representatives, Committee on Judiciary, "Ocean Freight Industry," Prepared by Antitrust Subcommittee, 87th Cong., 2nd sess., March 1, 1962 (Washington D.C.: Government Printing Office, 1962), p. 49.
8. Ferguson et al., *Economic Value,* pp. 170-171.
9. U.S. House of Representatives, Committee on Judiciary, "Ocean Freight Industry,'. p. 157.
10. *Ibid.,* pp. 159-169.
11. *Ibid.,* p. 186.
12. James R. Barker and Robert Brandwein, *The United States Merchant Marine in National Perspective* (Lexington, Mass.: Lexington Books, D.C. Heath and Company, 1970), p. I-13.
13. Ferguson et al., *Economic Value,* pp. 314-323.
14. *Ibid.,* p. 310.
15. Richard J. Morrison, "Moore-McCormack Lines, Inc. (A)." A case study at the Harvard Graduate School of Business Administration, ICH 9G13, AM-P, 163, 1963.
16. U.S. House of Representatives, Committee on Judiciary, "Ocean Freight Industry."

Chapter 3
Subsidized Ship Replacement Planning Process before 1961

1. National Academy of Sciences-National Research Council, "Trade-Fleet Study" (Washington, D.C., 1970), p. 10.

2. Allen R. Ferguson et al., *Economic Value of the Merchant Marine* (Evanston, Ill.: The Transportation Center at Northwestern University, 1961), p. 174.
3. Joseph T. Graziano, "Operating-Differential Subsidy Functions, Procedures, and Accomplishments" (New York: The Metropolitan Section of the Society of Naval Architects and Marine Engineers, March 19, 1964), p. 17. Published herein by permission of the aforementioned Society.
4. Herbert A. Simon, *Administrative Behavior* (New York: The Free Press, 1957), p. xxv. Copyright 1945, 1947, 1957 by Herbert A. Simon.
5. *Ibid.*, pp. 99–100.

Chapter 4
Environment of the Subsidized Lines: 1961 and Beyond

1. John L. Eyre, "Containerization: The Unhappy Marriage," *Dun's Review and Modern Industry*, June 1965, p. 135.
2. National Academy of Sciences - National Research Council (NAS - NRC), Maritime Cargo Transportation Conference, "Maritime Transportation of Unitized Cargo," Publication 745 (Washington, D.C., 1959).
3. Douglas C. MacMillan and T. B. Westfall, "Competitive General Cargo Ships" (New York: The Society of Naval Architects, Nov. 1960), pp. 845-847.
4. NAS-NRC, Maritime Cargo Transportation Conference, "Inland and Maritime Transportation of Unitized Cargo," Publication 1135 (Washington D.C., 1963).
5. U.S. House of Representatives, Committee on Merchant Marine and Fisheries, "Standardization of Containers," Hearings, 90th Congress, 1st sess. (Washington, D.C.: Government Printing Office, Oct.–Nov. 1967).
6. Testimony of James R. Barker on behalf of American President Lines, Ltd., and American Mail Line Ltd. before Federal Maritime Commission, Docket No. 69-56, April 1970, p. 14.
7. *Ibid.*, p. 16.
8. International Transport Workers' Federation, "Containerization" (London, 1968), p. 30.
9. Mckinsey and Co., "Containerization: The Key to Low-Cost Transport," sponsored by the British Transport Dock Boards (London, June 1967), p. 17.
10. Robert A. Hammond, "Emerging Changes in the Container Revolution," *Highway Research Board*, No. 281, p. 10.
11. U.S. House of Representatives, Committee on Merchant Marine and Fisheries, Subcommittee on Merchant Marine, "Construction Differential Subsidies-1966," 89th Cong., 2nd sess. (Washington, D.C.: Government Printing Office, April 1966), p. 33.
12. U.S. Department of Commerce, "MARAD 1969," (Washington, D.C.: Government Printing Office, 1970), pp. 82-83.

13. U.S. House of Representatives, Committee on Merchant Marine and Fisheries, "U.S. Maritime Policy" Hearings 90th Congress, 1st. sess. (Washington, D.C.: Government Printing Office, 1967), p. 13.
14. Henry S. Marcus, "Maritime Subsidy Board," p. 6. A case study at the Harvard Graduate School of Business Administration, 1970.

Chapter 5
Subsidized Ship Replacement Planning Process: 1961 and Beyond

1. Carl E. McDowell and Helen M. Gibbs, *Ocean Transportation* (New York: McGraw-Hill, 1954), p. 243.
2. The Propeller Club of the United States, "Proceedings of the American Merchant Marine Conference," Galveston, Texas, 1965, p. 69.
3. "Charter of Subsidized Ships for Gov't-Sponsored Cargo Viewed as Illegal by AMA," *Traffic World,* Nov. 22, 1969.
4. "Shipping: Storm Signals Lowering," *Dun's Review and Modern Industry,* May 1966, p. 125. Reprinted by special permission from DUN'S, May 1966. Copyright, 1966, Dun & Bradstreet Publications Corporation.
5. "R_X for the Merchant Marine," *Armed Forces Journal,* Feb. 15, 1969, pp. 29-30.
6. Litton Systems, "Oceanborne Shipping: Demand and Technology Forecast," June 1968, pp. 2-109.
7. *The Journal of Commerce, Library of Containerization,* Vol. 1 (New York, 1966), p. 5.
8. *Ibid.,* p. 5.
9. "SS Mormacsea, Versatile Ro/Ro Containerships," *Marine Engineering/Log,* May 1969, p. 48.
10. Equity Research Assoc., "Containerization," New York, May 16, 1969, p. 16.
11. *The Journal of Commerce, Library of Containerization,* Vol. 1, p. 30.
12. The Port of New York Authority, "Container Shipping: Full Ahead," New York, May 1967, pp. 24-25.
13. Equity Research Associates, "Containerization," New York, Dec. 15, 1969, p. 2.
14. The Port of New York Authority, "Container Shipping," pp. 24-25.
15. U.S. House of Representatives, Committee on Merchant Marine and Fisheries, "Cargo Container Dimensions," Prepared by Subcommittee on Merchant Marine, 90th Cong., 1st sess., Oct.-Nov., 1967 (Washington, D.C.: Government Printing Office, 1968), p. 128.
16. National Academy of Sciences-National Research Council, Maritime Cargo Transportation Conference, "Maritime Transportation of Unitized Cargo," Publication 745, (Washington, D.C., 1959) and "Inland and Maritime Transportation of Unitized Cargo," Publication 1135, (Washington, D.C., 1963).

Chapter 6
Environment and Fleet Improvement Planning Processes of Sea-Land Services and Matson Navigation Company

1. Wytze Gorter, *United States Shipping Policy* (New York: Harper and Bros., for the Council on Foreign Relations, 1956), p. 131.
2. *Ibid.,* p. 135.
3. *Ibid.,* p. 135.
4. Henry S. Marcus, "Sea-Land Service (A)." A case study for the Harvard Graduate School of Business Administration. ICH 14G29, 1970.
5. U.S. House of Representatives, Committee on Merchant Marine and Fisheries, "Standardization of Containers," Hearings, 90th Congress, 1st sess. (Washington, D.C.: Government Printing Office, Oct.-Nov. 1967).

Chapter 7
Analysis of Planning Processes: Subsidized and Unsubsidized

1. "The Box Worth Billions," *Forbes,* April 1, 1968.
2. NAS-NRC, "Trade-Fleet Study" (Washington, D.C., 1960), p. 10.
3. A. Scheffer Lang, "Innovation as a Coordination Requirement," in Marvin L. Fair and Howard C. Reese, eds., *Coordinated Transportation* (Cambridge, Md.: © 1969, Cornell Maritime Press, Inc.), p. 4.
4. U.S. House of Representatives, Committee on Merchant Marine and Fisheries, "Standardization of Containers," p. 150.
5. Leonard Baker, *The Guaranteed Society* (New York: MacMillan, 1968), p. 87.
6. Maitland S. Pennington, "Coordinated Transport and the Federal Maritime Acts," in Marvin L. Fair and Howard C. Reese, eds., *Coordinated Transportation,* © 1969, Cornell Maritime Press, Inc., Cambridge, Maryland, p. 115.
7. American Merchant Marine Institute, "A New National Maritime Policy," March 10, 1966, in U.S. Senate, Committee on Commerce, "U.S. Maritime Policy," p. 747.
8. U.S. House of Representatives, Committee on Merchant Marine and Fisheries, "Long-Range Maritime Program," 90th Cong., 2nd sess., April-May, 1968 (Washington D.C.: Government Printing Office, 1968), p. 185.
9. "Speedy Conversion Pays Off for Owner," *Marine Engineering/Log 75,* no. 1 (Jan 1970), p. 66.
10. "The Role of the Federal Government in the Shipping Industry: U.S. Maritime Laws." A case study for Harvard Graduate School of Business Administration, Boston, Mass., ICH 9G11, AM-P 161, 1963, p. 30.
11. The Interagency Maritime Task Force, "The Merchant Marine in National Defense and Trade: A Policy and a Program." October 4, 1965, in U.S. Senate, Committee on Commerce, "U.S. Maritime Policy," p. 581.

12. U.S. Department of Commerce, "MARAD 1969" (Washington, D.C.: Government Printing Office, 1970), p. 3.

Chapter 9
Summary and Conclusions

1. Maritime Cargo Transportation Conference, NAS-NRC,/"Maritime Transportation of Unitized Cargo," Publication 745 (Washington, D.C., 1959).
2. NAS-NRC, Maritime Cargo Transportation Conference, "Inland and Maritime Transportation of Unitized Cargo," Publication 1135 (Washington, D.C., 1963).
3. U.S. House of Representatives, Committee on Merchant Marine and Fisheries, "Long-Range Maritime Program," p. 525.

Bibliography

American Merchant Marine Institute, Inc. "A New National Maritime Policy," March 10, 1966.

Anthony, Robert N. *Planning and Control Systems, A Framework For Analysis.* Division of Research, Harvard Graduate School of Business Administration, Boston, Mass., 1965.

Arthur D. Little, *User Manual for the MARAD Fleet Simulations.* Contract No. MA-2451. Cambridge, Mass., 1964.

Baker, Leonard. *The Guaranteed Society.* New York: MacMillan, 1968.

Barkan, Joseph G. "Spotlight on the Construction Differential Subsidy Program." The Metropolitan Section of the Society of Naval Architects and Marine Engineers, New York, March 19, 1963.

Barker, James R., and Robert Brandwein. *The United States Merchant Marine in National Perspective.* Lexington, Mass.: Lexington Books, (D.C. Heath and Company), 1970.

Benford, Harry, "On the Rational Section of Ship Size." The Society of Naval Architects and Marine Engineers, New York, 1967.

____________"Principles of Engineering Economy in Ship Design." The Society of Naval Architects and Marine Engineers, New York, 1963.

Booz-Allen Applied Research, Inc. *Forecast of U.S. Oceanborne Foreign Trade in Dry Bulk Commodities.* Bethesda, Md., 1969.

"The Box Worth Billions" *Forbes,* April 1, 1968.

"Charter of Subsidized Ships for Government-Sponsored Cargo Viewed as Illegal by AMA." *Traffic World,* Nov. 22, 1969.

Equity Research Associates. "Containerization." New York, Dec. 15, 1969.

____________"Containerization." New York, May 16, 1969.

Eyre, John L. "Containerization: The Unhappy Marriage," *Dun's Review and Modern Industry,* June 1965.

Fair, Marvin L., and E. Grosvenor Plowman eds. *Coordinated Transportation, Problems and Requirments.* Cambridge, Md.: Cornell Maritime Press, 1969.

Ferguson, Allen R., Eugene M. Lerner, John S. McGee, Walter Y. Oi, Leonard A. Rapping, and Stephen P. Sobotka. *The Economic Value of the United States Merchant Marine.* The Transportation Center at Northwestern University, Evanston, Ill., 1961.

"Fifty Largest Transportation Companies." *Fortune,* July 15, 1966, pp. 258-59; June 15, 1967, pp. 222-23; June 15, 1968, pp. 214-15; and May 15, 1969, pp. 196-97.

Gorter, Wytze. *United States Shipping Policy.* New York: Harper and Bros., 1956.

Graziano, Joseph T. "Operating-Differential Subsidy, Functions, Procedures, and Accomplishments." The Metropolitan Section of the Society of Naval Architects and Marine Engineers, New York, March 19, 1964.

Hammond, Robert A. "Emerging Changes in the Container Revolution." *Highway Research Record,* No. 281, 1969.

Heess, Fred. "*SS Mormacsea,* Versatile Ro/Ro Containership." *Marine Engineering/Log,* May 1969.

Hummel, C. Ronald. "The Role of the Federal Government in the Shipping Industry: U.S. Maritime Laws." A case study, ICH 9G11, AM-P 161, Harvard Graduate School of Business Administration, Boston, Mass., 1963.

International Transport Workers' Federation. *Containerization.* London, 1968.

The Journal of Commerce. Library of Containerization, Vol 1. New York, N.Y., May 9, 1966.

Lawrence, Samuel A. *United States Merchant Shipping, Policies and Politics.* Washington, D.C.: The Brookings Institution, 1966.

Lewis, Edward V. "Research Toward More Efficient Transportation by Sea." The Society of Naval Architects and Marine Engineers, New York, N.Y., Nov. 1961.

Litton Systems, Inc. *Oceanborne Shipping: Demand and Technology Forecast.* Culver City, Cal. 1968.

MacMillan, Douglas C., and T.B. Westfall. "Competitive General Cargo Ships." The Society of Naval Architects and Marine Engineers, New York, Nov. 1960.

Marcus, Henry S. "American Mail Line, Ltd." A case study, T 686, Harvard Graduate School of Business Administration, Boston, Mass., 1969.

____________"Executive Lines (A)." A case study, ICH 14G30, Harvard Graduate School of Business Administration, Boston, Mass., 1970.

____________"Moore-McCormack Lines, Inc., Analysis of Roll-On/Roll-Off Ships," A case study, T 689, Harvard Graduate School of Business Administration, Boston, Mass., 1969.

____________"Maritime Subsidy Board." A case study, Harvard Graduate School of Business Administration, Boston, Mass., 1970.

____________"Sea-Land Service, Inc. (A). "A case study, ICH 14G28, Harvard Graduate School of Business Administration, Boston, Mass., 1970.

____________"Sea-Land Service, Inc. (B)." A case study, ICH 14G29, Harvard Graduate School of Business Administration, Boston, Mass., 1970.

"Maritime Hearings of House Subcommittee Focus on Operating Subsidy Proposals." *Traffic World,* Feb. 28, 1970.

McKinsey and Company. *Containerization: the Key to low-cost transport.* sponsored by the British Transport Dock Boards, London, June 1967.

McDowell, Carl E., and Helen M. Gibbs. *Ocean Transportation.* New York: McGraw-Hill, 1954.

Miller, Dave S. "The Economics of the Containership Sub-system." *Marine Technology* 7, no. 2 (April 1970): 180–95.

Morrison, Richard J. "Moore-McCormack Lines, Inc. (A)." A case study, ICH 9G13, AM-B.163, Harvard Graduate School of Business Administration, Boston, Mass., 1963.

Murr, Alfred. *Export/Import Traffic Management and Forwarding.* Cambridge, Md.: Cornell Maritime Press, 1967.

National Academy of Sciences - National Research Council, Maritime Cargo Transportation Conference. *Inland and Maritime Transportation of Unitized Cargo.* Publication 1135. Washington, D.C., 1963.

__________ *Maritime Transportation of Unitized Cargo.* Publication 745. Washington, D.C., 1959.

__________ *Research Techniques in Maritime Transportation.* Publication 720. Washington, D.C., 1959.

__________ *Trade-Fleet Study.* Washington, D.C., 1960.

Operations Research Society of America. *Operations Research* 6, no. 5 (Sept.-Oct. 1958).

Planning Research Corporation. *Transoceanic Cargo Study.* Los Angeles, Calif., 1971.

The Port of New York Authority. *Container Shipping: Full Speed Ahead.* New York, May 1967.

Propeller Club of the United States. *Proceedings of the American Merchant Marine Conference.* Galveston, Texas, 1965.

Reese, Howard C. and Marvin L. Fair, eds. *Merchant Marine Policy.* Cambridge, Md.: Cornell Maritime Press, 1963.

"Shipping: Storm Signals Lowering?" *Dun's Review and Modern Industry,* May 1966.

Simon, Herbert A. *Administrative Behavior.* New York: Free Press, 1957.

"Speedy Conversion Pays Off For Owner." *Marine Engineering/Log* 75, no. 1 (Jan. 1970).

Stanford Research Institute. *Projections of Principle U.S. Dry Bulk Commodity Seaborne Imports and Exports for 1975 and 1995.* Menlo Park, Cal., 1969.

"Testimony of James R. Barker on Behalf of American President Lines, Ltd. and American Mail Line, Ltd. Before the Federal Maritime Commission." Docket No. 69-56, April 1970.

U.S. House of Representatives, Committee on the Judiciary Antitrust Subcommittee. *Ocean Freight Industry.* 87th Congress, 2nd Sess., March 1, 1962, Washington, D.C.: Government Printing Office, 1962.

U.S. House of Representatives, Committee on Merchant Marine and Fisheries. "Maritime Authorization - 1970," 91st Congress, 1st sess., May 9, 1969. Washington, D.C.: Government Printing Office, 1969.

__________ "Cargo Container Dimensions," Hearings, 90th Congress. 1st sess., Oct.-Nov. 1967. Washington, D.C.: Government Printing Office, 1968.

__________ "Constructional Differential Subsidies-1966," 89th Congress, 2nd sess., April 1966. Washington, D.C.: Government Printing Office, 1966.

__________ "Long-Range Maritime Program." 90th Congress, 2nd.

sess., April-May, 1968. Washington, D.C.: Government Printing Office, 1968.

U.S. Senate, Committee on Commerce, Subcommittee on Merchant Marine and Fisheries. "Standardization of Containers." 90th Congress, 1st, sess. Washington D.C.: Government Printing Office, 1967.

____________"U.S. Maritime Policy." 90th Congress, 1st. sess., April-Aug. 1967. Washington, D.C.: Government Printing Office, 1967.

U.S. Department of Commerce. "Annual Report of the Maritime Administration 1968." Washington, D.C.: Government Printing Office, 1968.

____________MARAD 1969, *Year of Challenge: The Annual Report of the Maritime Administration for Fiscal Year 1969*. Washington, D.C.: Government Printing Office, 1970.

____________MARAD *1970, Year of Transition: The Annual Report of the Maritime Administration for Fiscal Year 1970.* Washington, D.C.: Government Printing Office, 1971.

____________MARAD *1971, Year of Breakthrough:* The Annual Report of the Maritime Administration for Fiscal Year 1971. Washington, D.C.: Government Printing Office, 1971.

____________MARAD *1972,* A New Wave in American Shipping: The Annual Report of the Maritime Administration for Fiscal Year 1972. Washington, D.C.: Government Printing Office, 1972.

U.S. Interagency Maritime Task Force. "The Merchant Marine in National Defense and Trade, A Policy and a Program," October 4, 1965.

Will, Admiral John M. "R_X For The Merchant Marine." *Armed Forces Journal,* Feb. 15, 1969.

Woodward, J. B., III, Benford, Harry, and Nowacki, Horst. "Systems Analysis in Marine Transport." The Society of Naval Architects and Marine Engineers, New York, 1968.

Index

About the Author

Henry S. Marcus is currently at the Massachusetts Institute of Technology in the positions of assistant professor of Marine Systems in the Department of Ocean Engineering and executive officer of the Commodity Transportation and Economic Development Laboratory. Dr. Marcus' experience includes consulting on a wide variety of problems in the area of transportation and logistics. He holds the D.B.A. from Harvard Graduate School of Business Administration, two M.S. degrees from Massachusetts Institute of Technology, and the B.S. degree from Webb Institute of Naval Architecture.